Real Estate Financial Modelling in Excel

Get ahead of your peers with *Real Estate Financial Modelling in Excel*, a book specifically designed to ensure that the next generation of property professionals become experts in the quantitative analysis of investments by teaching them how to create automated spreadsheets for the analysis of risk and return.

Real estate financial modelling has become an essential skill to investment analysts as the global property industry has seen huge transformations as a result of more institutional investors, especially private equity funds, increasing their interest in the asset class. Consequently, the industry requires a new skill set from real estate professionals and graduates.

Real Estate Financial Modelling in Excel will help current finance and real estate students, as well as practitioners, to harness the power of Microsoft Excel in the context of real estate investments and explain in an easy-to-follow manual style how to create financial models that will predict financial returns and the risks related to them. Readers will learn to use Excel for automation, data analysis, and data visualisation to inform their capital allocation decisions, giving them the edge with those technical skills in high demand in the investment markets and in particular with sophisticated investors such as pension and insurance funds, private equity, and specialised debt funds.

This book will address the needs of busy real estate professionals and students in the final year of a real estate bachelor's degree or master's degree, who want to apply the theories of finance and investment into practice and build models to help make decisions regarding acquisitions, disposals, and management of real estate assets.

Maria Wiedner is the Founder and CEO of Cambridge Finance, a company she founded in 2016 to provide real estate financial modelling courses and consultancy. She is also the Founder and CEO of Real Estate Women, CIC, an organisation which advocates for equality and inclusion and the fair representation of women and minorities in senior positions. In addition, Maria serves as a Real Estate Finance Specialist and APC Assessor for the Royal Institution of Chartered Surveyors (RICS). Maria currently teaches Real Estate Modelling in the Global Masters in Real Estate Development programme at IE School of Architecture and Design in Madrid, Spain, and is a D&I Editorial Advisor for *Property Week*, the leading news magazine in the commercial and residential property market in the United Kingdom. Maria's career has been built upon highly quantitative roles, such as property derivatives, renewable energy financial analysis, and listed real estate investment trusts (REITs) research. She is a graduate of Fundacao Getulio Vargas in Sao Paulo, Brazil, where she received a BA in Business Administration, an alumna of the University of Cambridge (MPhil), a CFA Charterholder, and a member of the RICS (MRICS). Having lived in Brazil (her home country), Finland, Austria, the United States, France, and the United Kingdom, Maria speaks Portuguese, Spanish, German, Finnish, and English.

Real Estate Financial Modelling in Excel

Maria Wiedner

Routledge
Taylor & Francis Group
LONDON AND NEW YORK

Designed cover image: © Shutterstock

First published 2025
by Routledge
4 Park Square, Milton Park, Abingdon, Oxon OX14 4RN

and by Routledge
605 Third Avenue, New York, NY 10158

Routledge is an imprint of the Taylor & Francis Group, an informa business

© 2025 Maria Wiedner

The right of Maria Wiedner to be identified as author of this work has been asserted in accordance with sections 77 and 78 of the Copyright, Designs and Patents Act 1988.

All rights reserved. No part of this book may be reprinted or reproduced or utilised in any form or by any electronic, mechanical, or other means, now known or hereafter invented, including photocopying and recording, or in any information storage or retrieval system, without permission in writing from the publishers.

Trademark notice: Product or corporate names may be trademarks or registered trademarks, and are used only for identification and explanation without intent to infringe.

British Library Cataloguing-in-Publication Data
A catalogue record for this book is available from the British Library

ISBN: 978-1-032-54443-4 (hbk)
ISBN: 978-1-032-54400-7 (pbk)
ISBN: 978-1-003-42489-5 (ebk)

DOI: 10.1201/9781003424895

Typeset in Times New Roman
by codeMantra

Contents

Acknowledgements	*vi*
Preface	*vii*

SECTION 1
Real Estate Financial Modelling – The Property and Its Income-Generation Potential — 1

1 What Are Financial Models and How Are They Used in the Real Estate Sector?	2
2 Real Estate Valuation Terminologies	10
3 Real Estate Valuation Using Capitalisation Methods	16
4 Introduction to Discounted Cash Flow Modelling and Analysis	43
5 Advanced Discounted Cash Flow Modelling and Analysis	63

SECTION 2
Financial Modelling of Commercial Real Estate Debt – Gearing (or Leveraging) — 103

6 Debt Structures	104
7 Comparing Debt Structures and Analysing Results	149

SECTION 3
Real Estate Development Financial Modelling — 165

8 Development Valuation and Analysis	166
9 Geared Development Investment	184

SECTION 4
Risk Modelling — 195

10 Sensitivity Analysis	196
Index	*229*

Acknowledgements

First and foremost, I would like to thank my family – my husband Rene and daughters Isabela and Sophia, for being my all-time supporters, inspiring me to be a better person every day.

I also want to thank my colleagues, who I also call family, at Cambridge Finance: Anouk Khan, Holly Mapletoft, and Jamil Al-Jawish. They were fundamental in keeping me sane and the business afloat. Thank you for allowing me to follow my passion in teaching and advising my clients on this very subject matter of real estate finance and investment.

To my clients and students: I wrote this book because you asked me to step up in my knowledge curve and deliver something better every day. You taught me, as much as I taught you.

To those who helped me make this book a reality: Ed Needle and Martha Luke, the very supportive editors who believed in my book ideas and followed up on every question this novice writer had to ask. I also want to thank Maan Al Maddah, Sherry Yishuang Xu, and Neil Crosby who provided the initial feedback on the manuscript and supported the application of this book. I also want to thank Michael Dunning who allowed me to use one of his properties as a case study in this book and those who helped get specific permissions to publish copyrighted materials: Alfonzina Celindano at FTSE Russell, Jackie Bowie at Chatham Financials, Charles Golding at the RICS, and more generally Trading Economics, Bank of England, OECD, and Statista, all of which provided me with some crucial data to be used in this book.

Last but not least, to our new readers (you!): Thank you for your interest! I hope you enjoy reading this book, and, most importantly, that you learn new fundamental skills that should carry you through your real estate investment career progression.

Preface

This book is the result of my professional experience in the real estate finance industry as a financial analyst, building models, but most importantly, using the models that I have created to provide reasoned advice to property investors and operators.

Over my 15 years of experience in real estate, I have built hundreds of models from the smallest of developments, such as single-family residential units up to mega projects, such as whole neighbourhoods which included many commercial and residential properties.

Very often, people ask me why they should learn financial modelling when they can create their models without all the complicated formulas, as they add a lot of 'inputted' numbers, or technically speaking, 'hardcoded' values. My answer is that a financial model is like a good construction: if you build it right in the first place, then you will have very little snagging issues and costs as you will build a robust product, whereas if you do not build it right, you will have more problems and costs along the way.

Another question that I always hear is whether building models should be left to those number crunchers. My answer is always that whereas I believe people have different skills, you should at least have a basic understanding of finance, financial mathematics, and modelling if you are going to play the real estate investment game. This book should serve as both a guide to those who want an 'eagle-eye' view of what financial modelling for real estate investment is and a handbook to those who want to build the models themselves.

As we are in the era of AI (artificial intelligence), another question that I often get is why I am still using Excel when the world has evolved to AI and we could model in Python, Java, etc. The reason for keeping my models in Excel (and I have further explanation in this book) is that real estate investments are very often cumbersome in the way they operate, and the way investment returns are derived is all quite different. For example, investing in a single-family unit without a mortgage is massively different to investing in commercial property with a mortgage with further refurbishment needed and financed by a non-collateralised loan. As such, in order to be both adaptive and creative, Excel seems to be the most flexible of all calculation tools, hence my application and language of choice.

Whilst existing software (some in Excel included) will calculate financial returns and values for real estate assets without needing to create or even understand a single formula are available, if you want to be one step ahead of the competition, being able to build financial models will certainly set you apart and give you a competitive edge in the world of real estate finance and investment as you will be able to be more creative and knowledgeable than your peers.

The models I present in this book were built from scratch, benefitting from a lot of feedback from clients, colleagues, and students. Although I am not acknowledging them all individually here, any comments that I have received about my models were noted and were reflected in future iterations. As a shout out, if you – along the way of reading and implementing the models – find any areas of improvement in my assumptions or errors in my models, I would love to hear from you.

Practical Information about the Book:

Who Am I Writing This Book For?

This book will suit those who want to build financial models or understand how they are built and the mechanics behind them. As such, this book can be used by students undertaking real estate degrees – both in a bachelor's or master's degree – or those studying finance, economics, or business administration with an interest in real estate. Moreover, this book is also for practitioners in the real estate industry who are already working as an investment analyst or wanting to make a career move to more investment-related jobs.

What Kind of Excel Knowledge Do You Need to Follow This Book?

Even though I will explain all the rationale and Excel formulas from scratch, it will make your life a lot easier if you already have some basic Excel knowledge. For example, if you are familiar with at least the basic functions such as SUM, AND, OR and, very importantly, the IF functions, that will give you a head start. I will also show you some Excel shortcuts, but if you already know that CTRL + C is to copy and CTRL + V is to paste, then it means that you are already aware of keyboard shortcuts, and this will help you to make the transition to more advanced shortcuts.

What Is the Expected Prior Knowledge to Make the Most of This Book?

This book is about the practical application of real estate finance theory, including valuation. As such, it is expected that readers should have a good understanding of these theoretical concepts, especially regarding discounting cash flows and some familiarity with property valuation terminologies such as yields (cap rates), lease events, and rents – for example, passing rent, estimated rental value, headline rent, etc.

I also recommend that readers are familiar with the RICS Global DCF Guidance Note[1] published in November 2023, to which I also contributed as a member of the Expert Working Group that formulated the Guidance. I also recommend that readers have read or are familiar with the concepts in established real estate valuation and investment books such as *Property Investment Appraisal* (2021)[2] by Baum A., Crosby N. and Devaney, S. and *Real Estate Investment* (2023)[3] by Baum, A.

Will the Excel Models Be Available to the Readers?

Yes, the models and additional resources, such as videos and further reading materials, will be available on our website www.cambridgerefinance.com.

Notes

1 Discounted Cash Flow Valuations, RICS practice information, global, 1st edition, November 2023.
2 Baum, A., Crosby N., Devaney S. (2021) *Property Investment Appraisal*, 4th Edition.
3 Baum, A. (2023) *Real Estate Investment: A Strategic Approach*, 4th Edition.

Section 1

Real Estate Financial Modelling – The Property and Its Income-Generation Potential

Real estate financial modelling is a vital component in the analysis and valuation of real estate investments. The first session of this book focuses on the asset side of the equation, which means looking at the property and its income-generation potential, assuming 100% equity investments in real estate. We will explore the fundamental principles and methodologies used to evaluate real estate assets, understand their potential returns, and assess their risks. The objective is to provide a comprehensive guide to modelling the asset side, with a particular emphasis on the practical aspects of real estate financial analysis and we will provide you with real-life case studies for practical applications.

We will first start with the principles of financial modelling, what it is and what it isn't, golden rules and how to apply best practices for building capitalisation and cash flow models for the valuation of single-let and multi-let properties.

Chapter 1

What Are Financial Models and How Are They Used in the Real Estate Sector?

Chapter Contents

Introduction	3
What Is Financial Modelling?	3
Uses of Financial Model	3
Investment Analysis	3
Risk Assessment	3
Financing	3
Budgeting and Planning	4
Valuation	4
Decision Making	4
Strategic Asset Management	4
Transparency and Communication	4
Market Analysis	4
Benchmarking	4
Why Excel?	5
Virtually Free	5
Auditable and Transparent	5
Flexible	5
Slowness	5
Errors	5
Best Practices in Financial Models	5
Phase 1: Understand the Business Case	6
Phase 2: Create the Layout	6
Phase 3: Develop the Formulas	6
Phase 4: Revise Model Structure	6
Phase 5: Rigorously Test the Model	7
Phase 6: Documentation	7
Golden Rules for Financial Models	8
Rule No. 1: Simple and Elegant	8
Rule No. 2: ONLY One Formula per Row or Column	8
Rule No. 3: USE Intermediate Calculations	8
Rule No. 4: NEVER Use Hardcoded Values in Any Formulas	8
Rule No. 5: Add Sensitivity Tables	8
Conclusion	9

This chapter also provides best practice recommendations for building a financial model from scratch in Excel and golden rules that will make a real estate financial model robust, easy to read, audit, and use in a simple and elegant way.

By providing a framework to develop financial models in the real estate sector, this chapter equips readers with the knowledge to understand the dynamics of real estate financial modelling and the confidence to start applying these principles in practice.

Introduction

What Is Financial Modelling?

Financial modelling means creating a spreadsheet, which is used to forecast an asset performance and help us make financial decisions. Throughout this book, we will create a financial model using Microsoft Excel to produce the valuation and analysis of a property. This valuation will then help us decide if we are going to buy, hold or sell the property and the implications of each decision as they represent their unique paths using their own risk and reward potential.

By using detailed financial models, real estate investors can make rational decisions, without relying too much on 'gut feeling' or 'rules of thumb' as their analysis will be more aligned with their risk and return targets. This is of particular importance when investment managers have fiduciary duty to retail investors, such as pension and insurance fund managers, as their ability to communicate their investment rationale is paramount to their jobs.

As such, financial models should form the backbone of investment decisions, as they can provide a robust and transparent way to analyse the key drivers of returns and composition of value, and highlight the areas of risks. Consequently, financial modelling plays a crucial role in the real estate sector, providing both professionals and investors with valuable insights to make informed decisions.

Uses of Financial Model

Financial models can help us with:

Investment Analysis

Financial models help investors understand the potential returns from a real estate investment. By forecasting cash flows, expenses, and potential changes in value, investors can determine the profitability of an investment opportunity.

Risk Assessment

A good financial model allows for sensitivity, scenario analysis, stress testing and simulations, where various risk factors are adjusted to see how they impact the project's financials. This might include changes in interest rates, occupancy rates, or construction costs, among others. This helps in understanding the sensitivity of the investment to various factors and to plan accordingly.

Financing

When seeking financing for real estate projects – both debt and equity – whether it's from banks, private investors, or other sources, a robust financial model is often required. Lenders and equity providers will want to see projected cash flows, debt service coverage ratios, and other financial metrics to assess the creditworthiness of the project.

Budgeting and Planning

For developers, financial modelling is critical for budgeting and planning purposes. It helps in estimating the costs associated with development, understanding the timeline of cash inflows and outflows, and ensuring the project remains financially viable from start to finish.

For asset managers, financial modelling is a vital tool to understand and control the budget associated with the investment, such as refurbishment, operating expenses, as well as expected rental inflows and lease events, such as expiries, breaks, renewals, reviews to market rents, and indexation.

Valuation

Financial models are essential tools in real estate valuation. Using techniques like discounted cash flow (DCF) analysis, investors and professionals can estimate the present value of future cash flows to determine the value of a property or portfolio.

Decision Making

Real estate professionals often face decisions like whether to buy or lease, renovate an existing property, or sell an asset. Financial modelling provides quantitative insights that can guide these decisions, ensuring they are based on sound financial reasoning.

Strategic Asset Management

For those managing a portfolio of real estate assets, financial modelling can assist in determining optimal strategies for asset management. This might include decisions about property upgrades, lease renewals, or sale.

Transparency and Communication

A well-constructed financial model provides transparency. For teams, stakeholders, or potential investors, it serves as a clear representation of the assumptions and expected outcomes of a real estate project. This aids in effective communication and alignment of expectations.

Market Analysis

By incorporating market research and trends into financial models, real estate professionals can forecast how shifts in the market, such as changes in demand or rental rates, might impact a project's profitability.

Benchmarking

Financial models can be used to benchmark a project or investment against industry standards or comparable projects to ensure competitiveness and viability.

In summary, financial modelling in real estate serves as a compass, guiding investors and professionals through the complexities of the sector. It provides a structured way to analyse, visualise, and understand the financial implications of decisions, ensuring that risks are mitigated and opportunities are maximised.

Why Excel?

As the main spreadsheet tool these days, Excel is a great application to automate calculations. It is a virtually free, auditable, transparent, and flexible.

Virtually Free

If you subscribe to the Microsoft Office package, the full Excel version is included, and these days, it is impossible to be a computer user without subscribing to a Microsoft Office package.

Auditable and Transparent

All calculations in Excel can be traced back. If sheets and formulas are hidden and protected, this can be disclosed in the form of providing the password. Moreover, the Excel library is very well understood and accepted as market standards.

Flexible

In Excel, you can change any formulas – if they have not been locked by a password – and make it adaptable to the scenarios you want to model. For example, if you want to change some features in the lease structures, you can perhaps add a new 'IF' function to reflect this new scenario.

However, Excel does not come without its pitfalls, such as slowness and increased margin for error if left unprotected.

Slowness

If there is one major problem in using Excel, it is how slow it can get – mainly when using a lot of data tables for sensitivity analysis; too many formulas using ARRAYS or volatile functions, i.e. those functions that will make the whole spreadsheet to recalculate even if the arguments do not such. Examples of volatile functions are OFFSET, INDIRECT, TODAY, NOW.

Errors

Excel is an amazing tool that will allow you to be creative in designing financial products; however, it is very easy to make mistakes. One bracket (or lack thereof) can give you a completely wrong answer, and it can be difficult to find the error if you are not a skilful modeller.

Best Practices in Financial Models

Spreadsheets for financial modelling are a critical element within complex business processes such as business planning and regulatory reporting as described before. In real estate, it is no different. More and more property professionals are relying on Excel or other software to provide valuations, investment, divestment advice or risk assessment. In the absence of a generally accepted and widely applied standard, Excel developers build models according to their own tastes, but by using some basic standards, you should be able to create a model that is consistent and improves the overall efficiency of your work. It is important to note that no single financial model will be right for every property investment analysis. Property is by definition unique and fine-tuning the financial model according to the peculiarities of each property, and transaction, is fundamental.

Below we will describe the phases of creating a financial model from scratch. By following those phases and using our recommendations, this will produce a financial model that will incorporate best practice guidelines.

Phase 1: Understand the Business Case

Understanding the business case is crucial when developing your financial models, as the more you understand the project's vision and decision-making framework, the more fit-for-purpose your model will be. For instance, if you are building a financial model for purchasing a prime office fully occupied by Apple or Microsoft, the model and risk-return assessment will be very different from that of an investment in a shopping centre in a secondary market. Consequently, by grasping the location's economics, the structure of tenancy agreements, and the building's condition, you can create a model that is far more effective in supporting the end-user's (investor's) financial decisions.

Together with the end-user, you will need to identify the inputs and outputs necessary for the financial model, so that the spreadsheet has a clear purpose and defined output.

Recommendations:

- Do not start to create a financial model before satisfying yourself that you understood the scope of the job, i.e. you need to understand the business plan
- Work collaboratively with users, clearly defining the inputs and outputs of the model

Phase 2: Create the Layout

Now that you know for what the business plan will be – for example, buying land, building a shopping centre, finding the tenants, holding for five years, and selling it – it is time to create the layout, i.e. defining the inputs, outputs, and the method of calculation. Models should be laid out in such a way that users can understand the flow of logic and modify it as appropriate. It should be like a book, with a story to tell, starting with an introduction, developing the narrative, and ending with a conclusion. It's important to remember that the model should always look clean, organised, and following a certain order.

Recommendations:

- Do not start creating cash flows, balance sheets, and asset valuations unless you have a layout and structure in front of you
- Clearly separate inputs, workings, and outputs. If possible, keep inputs and outputs on the same sheet, so you can check the impact of input changes in outputs
- **Keep the financial model simple and elegant!**

Phase 3: Develop the Formulas

Now that the layout has been devised, roll up your sleeves because building formulas that will work under several scenarios is what the financial modelling profession is all about. It is very unlikely that you will get it right on the first attempt or that the formula will be the simplest and most efficient one. But the idea is that you should only start creating formulas once you have understood the business plan, set up a layout that is simple and elegant and have decided on the best method that will link the inputs to the outputs in the most efficient manner.

Phase 4: Revise Model Structure

An important feature of a financial model is how user-friendly and transparent it is. A model is not static, users will interact and manipulate it constantly. Creating a model that is easy to use and understand is a complex task, but we should aim for it.

Once you have created the formulas and linked all the inputs with the outputs, you need to check again to see if there is something else that you could improve in it. For example, a very complex formula, duplicate input data, or an input data not being clearly identified.

If you have carefully followed Phase 2, your model should already have a coherent structure, but as we come to the end of the modelling process, it is good practice to review it and check if you have adopted best practice guidelines. Below are some recommendations to make sure your model has a structure that is user-friendly and transparent.

Recommendations:

- Ensure that the model structure follows the rule of simplicity and elegance
- Use data validation if the input data is constrained by a small number of options
- For example, a user can select between five (5) types of loan repayment only
- Format it in a logical way

Formatting is essential in any spreadsheet. This not only helps the user to make sense of what they need to do but also make it more appealing to non-users, such as investors and other stakeholders looking at the model.

Below are some formatting rules that you should follow:

- Differentiate clearly input cells from outputs. Use background colours for input cells so users know which data can be changed in the model
- Use conditional formatting to highlight relevant changes (for instance, change in rents)
- Build-in error checks and apply conditional formatting in order to catch the user's attention
- Avoid referencing cells in one workbook from another. If you *must* spread your workbook across multiple workbooks, make it obvious which values are linked to and from other workbooks. For example, host all such links in a 'landing page' contained in the model and create a check that links have been updated
- Remove gridlines after you finish your model
- Pre-format pages for easy printing
- Protect your worksheet and check only input cells can be edited but not output cells
- Don't forget to create a password log for different worksheets and workbooks

Phase 5: Rigorously Test the Model

Testing is a vital step to identify and eliminate errors.

If you have built error checks in your model, you should be able to pick up errors early in the process, but you still have to find out if the model is fit for the purpose, i.e. if it works under different scenarios.

Recommendations:

- Add sensitivity analysis and check if results make sense
- For example, if you change sale price alone, will this have the expected impact on returns, i.e. price up, return up?
- Test different data and see if the model copes with it. For example, change dates, rents, and yields and check if the result is as expected
- Ask a colleague to test it
- Consider commissioning an independent company to audit your model

Phase 6: Documentation

It is important to write the assumptions and limitations of the model. It can be a separate document in Word, but it can also be a sheet in your financial model. At a minimum, add a description page, main assumptions and date it was last updated and by whom. You should also include the source of any forecast figures, such as yields, market growth, and inflation if using third-party or house figures.

Golden Rules for Financial Models

These are my tips, but I will refer to them as 'golden rules' since following them have saved a lot of headache and avoided errors, and I *always* have them in mind when developing my models:

Rule No. 1: Simple and Elegant

The simpler model appearance and the formula, the better.

Sometimes, modellers have hard rules, such as 'no formulas longer than your thumb' and 'no line breaks', but I prefer to leave this to the modeller to decide. The trade-off of very simple formulas is that we will then need a much larger workbook with several tables. Whereas if we add a bit more complexity to the formulas, we will end up with a workbook that is visually simpler and more elegant, (I prefer the latter).

The only hard rule I have is that no formulas should have more than **three (3) IF functions or IFS conditions**.

Rule No. 2: ONLY One Formula per Row or Column

The idea here is that you will need to create one formula that will deal with all scenarios in future periods. This means that formulas need to be consistent throughout the model, so you just need to copy a cell from left to right, or top to bottom, without needing to change it.

Exception: you may need to use one single 'anchor' cell as the start of a series. It is okay. But from there, all formulas should be the same.

Rule No. 3: USE Intermediate Calculations

If you see your formula ending up with more than three IF functions or looking too complex, you should think of using intermediate calculations.

For example, you can use an intermediate dates calculation table to change all actual dates into cash flow dates or separate out the lease events and add them up in a total rents table afterwards.

Rule No. 4: NEVER Use Hardcoded Values in Any Formulas

Hardcoded values are fixed values embedded into a formula.

For example, let's say that you plug 2.0% as market growth rate to be calculated on a compounded basis as rent increase. If the growth rate forecast changes to say 2.5%, you will need to change manually all 2.0% entries. This will considerably increase error risk in your model. Instead, consider the growth rate as an input in your input table and refer all formulas to the input cell instead.

Rule No. 5: Add Sensitivity Tables

The sensitivity tables – which will be discussed in more depth further in the book – are part of the 'What-If' analysis and checks for the impact of changes in input variables on outputs. For example, what is the internal rate of return (IRR) if market growth changes from 2.0% per year to 2.5%, 3.0% and so on? This question can be answered with a sensitivity table.

The reason for the sensitivity is twofold: risk analysis ('what if') and to test if the model is working properly, because if there are hardcoded values, for example, the growth rate is hardcoded in rent formula, the sensitivity tables won't work and that will highlight an error in the financial model.

Conclusion

This chapter looked at best practice in financial modelling, gave you some recommendations to keep on track and some rules that you should follow when creating a financial model from scratch.

The important thing is to keep financial models simple, elegant, and error-free. This will come with practice and experience, but with these simple tips, you are on the right path to achieve it.

Chapter 2

Real Estate Valuation Terminologies

Chapter Contents

Introduction	10
Property Yields (or Cap Rates)	11
Initial Yield (or Going-In Cap Rate)	11
Reversionary Yield	11
Exit Yield (or Going-Out Cap Rate)	11
Equivalent Yield	11
Net vs Gross Yields	12
Rents	12
Passing Rent	12
Market Rent/ERV/MRV	12
Headline Rent	13
Topped-Up Passing Rent	13
Gross Rent	13
Net Rent	13
Lease Events	13
Lease Starts	13
Rent-Free Periods	13
Rent Reviews/Renewals	13
Breaks	14
Expiries/Voids	14
Step-Ups	14
Indexed Rents	14
Conclusion	14

Introduction

Before we move on to creating real estate financial models, I think it is crucial to clarify and demystify the terminologies associated with real estate valuation that will be used in our models. Real estate valuation and its associated financial model involve a complex array of terms and concepts that can often be misinterpreted.

Moreover, American and British terminologies get intertwined and there are also many nuances in the terminologies when used across countries, for example, yields and cap rates, net rents and so many more.

The aim of this chapter is therefore to list the main terms used in real estate valuation and investment analysis so that when we produce our financial models, they use consistent terminology and we can ensure clear communication among stakeholders based on our analysis.

Property Yields (or Cap Rates)

Broadly speaking, property yields are the relationship of income (rent) and the price or value of the property. It is the same as the capitalisation rate (cap rate), but for consistency, I will use the more British term 'yield' instead (as opposed to the more American term of 'cap rate').

Example: if a property is rented for 50,000 and its selling price is 1 million, then the income represents 5.00% of the price, so this 5.00% is what we call 'yield'.

Another confusing aspect of yields is that there will be input and output yields. Input yields are those used to calculate the value, whereas output yields are those calculated once the value has been ascertained.

Below, I define the further derivatives of the term yield and how they are applied in the context of real estate valuations.

Initial Yield (or Going-In Cap Rate)

The initial yield is simply the passing rent divided by the value of the property and can be derived from the fully let comparable and then adjusted to represent the security or riskiness of the term income. Typically, in a rising market, this security comes from the fact that the risk of the tenant defaulting (i.e., not paying the rent) is less as they are less likely to leave the premises while paying less than current market rent.

Initial yields can be both an input as well as an output yield. For example, you can use an initial yield of 5.00% to calculate the value, but in an under-rented property, the output initial yield may be less than 5.00% due to the effect of higher market rents.

Reversionary Yield

The reversionary yield is the yield applied at the reversion, i.e., when rents become market rents. The reversion in a property is riskier than the term of the lease. For example, a tenant may have a contract for five years paying 50,000, but the market rent is now 60,000; therefore, the yield applied to the term should be lower than the yield applied to the reversion because we do not know many aspects of the reversion, for example, the growth implied, the risk profile of the new tenant, and if indeed the market rent is 60,000 as this is a hypothetical value.

Exit Yield (or Going-Out Cap Rate)

The exit yield is used to calculate the terminal value, exit value, or sale price of the property. In the implicit method, this is not used because a terminal value is not calculated – only the reversionary value, but it is a fundamental input in a discounted cash flow (DCF) model.

Equivalent Yield

As an input yield, the equivalent yield is a single capitalisation rate that is applied to both current and expected market rent. The importance of the equivalent yield as an input yield is because valuers in practice will often simply apply the equivalent yield they think is appropriate (derived from market evidence and market sentiment) to the entire income stream to determine value, rather than trying to justify applying two different input yields to the income streams (especially when a risk premium applied to the reversionary income stream is very difficult to justify from evidence of transactions in the market).

As an output yield, the equivalent yield is calculated as the weighted average of the initial and reversionary input yields. It is weighted by time and money (rents).

As such, the equivalent yield gives the average running (year-on-year) yield of the property without the effect of growth (since both passing rent and market rents considered are as of the valuation date).

Net vs Gross Yields

All yields described above can be quoted on a 'net' or 'gross' basis. Net means that purchaser's costs were included in the price of the asset; therefore, it is lower than the gross yield. So, if you say that the price of the asset is 1 million, the passing rent is 100,000 and the purchaser's costs (stamp duty, broker's, and lawyer's fees) adds up to 10% of the asset price then:

Gross Yield = Rent/Purchase Price = 100,000/1,000,000 = 10.0%

Net Yield = Rent/(Purchase Price + Purchaser's Costs)
= 100,000/(1,000,000 × (1 + 10%)) = 9.09%

The net initial yield can be seen as the 'true day one return' on the property as it takes into account all costs involved in the transaction.

Some countries (UK, for example) will quote yields as net, whereas others may quote them as gross, and when using net yields, it is important to take out this cost (i.e., divide by (1 + purchaser's cost in %)) when determining an asset value.

Another important issue arising from the use of net yields is that you need to be mindful that if you are going in using a net yield, then you should apply the same logic to the going out cap rate or exit yield. If you just capitalise the exit rent by an exit yield without considering the effect of purchaser's costs in the yield, you will inevitably be overestimating your sale price.

Rents

Passing Rent

This is the current rent being received on the valuation or acquisition date. Note that sometimes the passing rent and the contracted rent may be different; for example, there may be a tenant in a rent-free period when the property is valued. Whether you should use the passing (zero) or the contracted rent will depend on the method of valuation. If using the implicit valuation, you may use the contracted rent, but if using a DCF, you will need to start with the rent only at the end of the rent-free period.

Market Rent/ERV/MRV

Market rent is the rental amount that the property – or unit within the property – would be leased for on the valuation date. This is very much based on past leasing transactions, for example: a new rent agreement in a comparable building within six months; a reviewed lease agreement to market rent; sometimes, the passing rent plus the estimated rental growth in the market (if there is some evidence thereof).

The official wording for market rent is

> the estimated amount for which an interest in real property should be leased on the valuation date between a willing lessor and a willing lessee on appropriate lease terms in an arm's length transaction, after property marketing and where the parties had each acted knowledgeably, prudently and without compulsion.
>
> (RICS Valuation Global Standards effective 31 January 2022)

Other terms for market rent are estimated rental value (ERV), open market rental value (OMRV), or market rental value (MRV).

Headline Rent

This is the contracted rent as it appears in the lease contract, without consideration of rent-free periods and other tenant incentives, such as capital, cash, or in-kind contributions (including tax contributions as capital allowances).

Topped-Up Passing Rent

The topped-up passing rent is the current rent (headline rent if used) plus the market rent (ERV/OMRV/MRV) of the vacant units. The reason for a topped-up passing rent is because most properties are sold on its economic value, given that the vacant units could be let at market rents or that the vendor guarantees a market rent for a certain period of time, normally one year, and this can be taken out of the net asset price.

Gross Rent

The gross rent is when the owner is responsible for paying for the maintenance, utilities, and insurance of the building, i.e. the service charge is not recoverable from the tenants. Typically, residential properties will be leased on a gross rent basis, whereas commercial properties, especially single-lets, will be leased on a net rent basis.

Net Rent

The definition of net rent that we will use in this book is that of FRI (full repairing and insuring in the UK) or triple net, meaning that the tenant is responsible for all maintenance, repair, insurance, and any property taxes. In the case of a stabilised asset (one fully let), the net rent and the net operating income (NOI) will be the same, as the owner will not have any extra costs of operating the building.

Lease Events

Lease Starts

If a property or unit is already occupied, the lease start will be prior to the acquisition or valuation date; therefore, it will not be relevant to the calculation as they are in the past. However, if the property or unit is vacant, then the lease start will be the assumed start date of a new lease considering the time to rent it up.

Rent-Free Periods

This is a form of tenant incentive, whereby the tenant does not pay rents for a certain period, but may pay service charges depending on the terms of the lease. This is a relevant lease event when the vacant unit will assumedly start paying rents or when the second, hypothetical tenant starts paying rent again after a period of refurbishment, vacancy, or void.

Rent Reviews/Renewals

If a property or unit therein is let, there will be a time when the rent will be renewed to market rents. This is the time when tenants and owners – or their representatives – will come to an agreement of the MRV and the new rent will be validated. In the UK, it tends to be in 5 (five) year cycles, but in other jurisdictions, it may be less (France, for example).

In case of short leases, such as in Saudi Arabia, where to date, commercial lease state agreements are for three or five years, the reviews can be considered as renewal events, where the owner and the tenant will agree a new rent.

Breaks

Breaks are option agreements in the contract whereby the tenant has the option to break the lease at an earlier date with or without penalties, or the owner of the building has the option to cancel the lease. They have become a common feature in the tenancy agreement in the past decade, mainly from tenant demand as it requires more flexibility on when to terminate the lease.

From an owner and lender's perspective, it does increase the risk of early termination, and values can be lower when tenants have an option to break earlier than expiry.

Having said that, not all breaks are exercised. For example, when rents grow faster than inflation because of market demand, and therefore, the property is under-rented or the cost of relocation to the tenant would be greater than any lower market rent. (Such as a purpose-built data centre where the relocation would incur further capital expenditure and operating costs).

Expiries

This is the end of the lease. In a long lease scenario (say 10 or 20 years), the tenant would be willing to relocate to a more modern premise. So, you would need to incur refurbishment costs and the cost of getting a new tenant. Once the lease expires, you will need to make assumptions for the period it will take you to find a new tenant (void period), and the costs incurred in securing the new tenant, such as tenant incentives (rent-free periods and fit-out cost contributions) as well as void costs and letting fees.

Step-Ups

Step-up rents are mainly found in retail properties, whereby the owner will start the lease at a much lower rent than the market rent and agree beforehand the increases in rent (for example, the market rent is 500,000 and the rent starts at 100,000 increasing to 500,000 in 3 years' time). Therefore, the rent agreement could be for 100,000 in year 1, 300,000 in year 2, and 500,000 in year 3. In the subsequent years, the rent could be adjusted to market rents depending on the negotiations.

Indexed Rents

Indexation typically refers to adjustment of rent to inflation benchmarks. This indexation tends to be annually, but in some jurisdictions, for example, in Germany, there may be clauses that the rent will only be adjusted if the cumulative inflation since the last indexation is greater than 10%. In Continental Europe, commercial rents tend to have a clause for indexation adjustments; hence, many real estate professionals argue that property is a good hedge to inflation; however, this may not be the case in countries where rents are fixed for a period of time – UK for example.

Conclusion

This chapter aimed at giving you a brief overview of the main terminologies used in real estate valuation and financial modelling in this book. Understanding and demystifying these terms should not be merely an academic exercise but a critical step in ensuring accuracy, clarity, and robustness in real estate model building and decision making.

The significance of this understanding is further amplified when we consider the variations in terminology between different regions, such as the United States, the United Kingdom, Middle East, and Continental Europe. For example, "passing rent", "market Rent", and "headline rent" might have specific nuances in different markets, being gross or net, and even the "net" terminology may be different but may mean the same instance.

Additionally, lease events – such as renewals, rent reviews, and break options – play a crucial role in shaping the cash flow and financial outcomes of real estate investments, and they are very much dependent on the legal framework and market practices in different countries. As such, it is the fundamental importance that real estate investment practitioners are well versed in the specific language of the market in which they operate and that their financial models reflect these nuances.

As we move forward in this book, this chapter should work as a glossary to guide your approach to real estate financial modelling, valuation and investment analysis.

Chapter 3

Real Estate Valuation Using Capitalisation Methods

Chapter Contents

Introduction	17
Under-Rented Scenario	17
Term & Reversion Method	17
Layer Method	18
Years Purchase for Under-Rented Scenarios	18
Term and Reversion vs Hardcore: Which One to Use?	26
Over-Rented	26
Term & Reversion	26
Layer Method	27
Years Purchase for Over-Rented Scenarios	27
Investment Summary	28
Rack Rented	29
All-Risks Yield Method or True Capitalisation Method	29
Output Yields	30
Initial Yield	31
Equivalent Yield	31
Steps	32
Yield Analysis and Net Value	34
What Can You Say about the Property Value Looking at the Yields Alone?	35
If You Assume Purchaser's Costs of 7.0%, What Would Be the Net Value of the Asset?	35
Formatting	35
Dates: Format a Number as Dates	36
Valuation Term: Format a Number to Include a Text Next to It	36
Yield: Format Percentages:	37
Passing Rent, YP and Values: Format with Thousands Separator and Adjust Decimal Places	37
Borders: Add Outside Borders	37
Changing Input and Output Cells Colours	38
Hiding the Gridlines	40
Protecting Sheet	41
Conclusion	42
Notes	42

Introduction

Capitalisation or implicit or traditional or conventional methods – yes, they can all be used interchangeably – refer to property valuations that use comparable market evidence of investment transactions to ascertain a property value. Capitalisation valuation techniques have been under attack for some decades now, and I am one of the most active advocates against it, having co-written a paper on Discounted Cash Flow (DCF) valuation for the Royal Institution of Chartered Surveyors (RICS).[1] However, as they continue to be a very common practice in property valuation, at least in the UK, I am adding this chapter.

Some of the more arduous advocates in favour of the implicit method argue that yields are more 'comparable', making this comparability ideal for market valuations. However, the problem with this is that each property is unique, more so the quality of tenants and lease structures; therefore, comparable yields are not that comparable and hide too many assumptions in one single figure. Understandably, this method is easy to grasp, so valuers stick to it.

Another problem with comparable yields is when markets are illiquid, i.e., transactions are rare, values may not even be reported, and market rents are more volatile or uncertain. In this case, using yields becomes almost irrelevant, but still, valuation practitioners stick to this method as they provide quick answers.

The capitalisation method can be broken down into three other sub-methods, such as:

- Initial yield – or all-risks yield – method
- Term & reversion
- Hardcore or layer method

In order to decide which method you should use, you need to look at the property being either under-rented, over-rented, or 'rack'-rented, and very importantly, the type of lease event that will determine the reversion to the market.

The purpose of this chapter is not to explain and delve into the theory of property valuation as there are some other textbooks that will do a much better job than me, but the idea of this chapter is to teach you how to do the calculations using Excel, i.e., building a financial model for doing a property valuation using the capitalisation method.

Under-Rented Scenario

Under-rented is the scenario in which passing rent is lower than current market rents or estimated rental value (ERV).

Because it is unlikely that the tenant will leave, this income portion is more certain and should be valued using a lower yield than the ERV.

How low it is will very much depend on the valuer.

Term & Reversion Method

In the Term & Reversion method, the income stream is sliced vertically and works well when used in a fully-let reversionary freehold scenario.

The main assumption is that the property is currently fully let, and the current estimated rental value (ERV) will be achieved when the tenancy agreement expires (reversion).

The diagram below shows an under-rented scenario, i.e., the contracted lease (passing rent) is at a value below current market rents (ERV).

Under-rented Term & Reversion Method

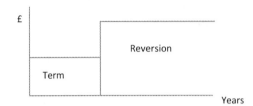

Figure 3.1 Under-Rented Term & Reversion Method.

In order to calculate the value of the property, we need to bring to the present value both the Term and the Reversion, thereby using a "shortcut" method of calculating these income streams. We will use the real estate jargon used to calculate this ratio: the years purchase (YP).

YP is, in fact, a present value ratio which when multiplied by the rents will result in the present values of each income stream.

Layer Method

The layer method, on the other hand, slices the income stream horizontally, and although there is no hard data on which methodology valuers use most, it does seem to be the most popular method currently. The top layer, in fact, shows the potential capital gain of the reversion and valuers tend to like this approach.

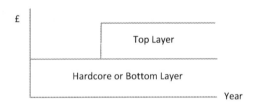

Figure 3.2 Under-Rented Hardcore Method.

Under-Rented Hardcore Method

The below table will give you all the formulas for YP needed to calculate the property value using the conventional valuation methods.

Years Purchase for Under-Rented Scenarios

	YP	Formula
Term	YP for n years	=(1-(1+yield)^-years) / yield
Reversion or Top Layer	YP in perp def n years	=(1+yield)^-years / yield *or* PV of 1/yield
Bottom Layer	YP in perp	=1/yield

Figure 3.3 Years Purchase for Under-Rented Scenarios.

Where *years* is the number of years to lease expiry and *yield* is derived from a comparable property yield.

Term & Reversion
CASE STUDY 1

17 Fleet Street, London EC4, Office Single-Let Midtown Investment

Figure 3.4 17 Fleet Street.

- Freehold
- Centrally located in the heart of London's legal district, less than 100 metres from the Royal Courts of Justice
- Excellent transport connections with five stations within 750 metres
- 6,082 sq ft (565.0 sq m) of office, retail, and ancillary accommodation arranged over lower ground, ground, and three upper floors.
- Single-let to Agencia per a la competitivitat de L'Empresa Accio, a public agency attached to the Ministry of Business and Employment of the Government of Catalonia, on a FRI (triple net) lease
- Passing rent of £215,000 per annum equating to a passing rent of £35.35 per sq ft, overall
- WAULT (weighted average unexpired lease term) of 4.6 years to expiry (from valuation date)

The valuation date is 31 December 2023, and the contract will expire on 12 August 2028. The ERV is £45 sq ft equating to an ERV of £273,690. The valuer has estimated a comparable equivalent yield of 8.0% and believes that initial yield should be 'adjusted' down to 6.0% given the security of the tenant and the fact that the asset is under-rented.

EXERCISE 1

Value this property using the Term & Reversion method.

To accomplish this, we will start from scratch:

First, set up the spreadsheet as below:

Figure 3.5 Term & Reversion Spreadsheet 1.

Inputs:

Start dates: when the passing rent and market rents will start.

Start date for passing rent: If the tenant is already in place and paying rent, then the start date will be the date of valuation. However, if this is a vacant unit or if the tenant is not paying rent yet due to rent-free period, then the start date will be in the future.

Start date for market rent: This should be the date of the lease event, i.e., review to market rent, renewal, break, or expiry. However, some valuers may also add a period of vacancy if the lease event is a break that is going to be exercised or an expiry. For example, if the lease expiry is on 30/06/2028 and the valuer estimates a one-year vacancy period, then it could be said that the start date would be 30/06/2029. This adjustment may not always be necessary, because generally vacancy allowance would be linked with the choice of yield, i.e., a 'softer' (higher) one. The latter is going to be our approach.

Valuation Term:

This is the duration of the applicable rent (passing or market rent) in the valuation. For the term, it will be the period up to the next lease event. Whereas for the market rent, it will be in perpetuity.

To calculate the valuation term for the term, we will use the function in Excel called YEAR-FRAC. This function gives the number of years between two dates, in our case, between the valuation date and the date of lease expiry.

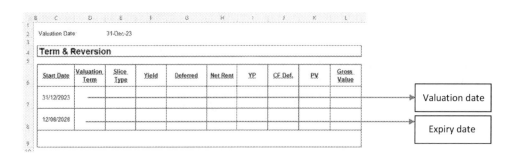

Figure 3.6 Term & Reversion Spreadsheet 2.

The syntax of YEARFRAC is:
= YEARFRAC (start date, end date, [basis])
Note that [basis] is optional and it is more relevant when calculating bonds or interest rates when day count conventions are necessary.

Start Date	Valuation Term	Slice Type	Yield
31/12/2023	=YEARFRAC(C7,C8)		
12/08/2028	In Perp Def		

Valuation Date: 31-Dec-23
Term & Reversion

Figure 3.7 Term & Reversion Spreadsheet 3.

We can further populate **Slice Type** and **Yield** as per descriptions in the case study.

Valuation Date: 31-Dec-23
Term & Reversion

Start Date	Valuation Term	Slice Type	Yield	Deferred
31/12/2023	4.6167 yrs	Term	6.00%	
12/08/2028	In Perp Def	Reversion	8.00%	

Figure 3.8 Term & Reversion Spreadsheet 4.

Deferred refers to the period between the valuation date and when the applicable rent (either passing or market) will start. For example, in the reversion, the market rent is deferred until the lease expiry and it is only then that the market rent will be accounted for. The way we will show it is in terms of years and months, and for this, we will use the function in Excel called DATEDIF.

The way the DATEDIF works is as follows:
=DATEDIF (start_date, end_date, "y" or "ym")

The "y" will show the number of years between the dates, and the "ym" will show the number of months between the dates. As such, we will need two DATEDIF if we want to show both the years and the months between the valuation date and the lease event date (in our case, the expiry of the lease).

We will also need to concatenate the text (using &) to add "yrs" and "mths" within the formula for DATEDIF. This is what it should look like:

=DATEDIF(valuation date, lease event date,"y")&" yrs "&DATEDIF(valuation date, lease event date,"","ym")&" mths"

	B	C	D	E	F	G
2	Valuation Date:		31-Dec-23			
4	**Term & Reversion**					
6		Start Date	Valuation Term	Slice Type	Yield	Deferred
7		31/12/2023	4.6167 yrs	Term	6.00%	=DATEDIF($E$2,C7,"y")&" yrs "&DATEDIF($E$2,C7,"ym")&" mths"
8		12/08/2028	In Perp Def	Reversion	8.00%	=DATEDIF($E$2,C8,"y")&" yrs "&DATEDIF($E$2,C8,"ym")&" mths"

Figure 3.9 Term & Reversion Spreadsheet 5.

Net rents were given in the description of the case study, passing rent = 215,000 and market rent = 273,690. If the rents were not given net, you would need to deduct the applicable operating expenses from them in order to achieve the net rent that will be capitalised by the yield.

	B	C	D	E	F	G	H	I
2	Valuation Date:		31-Dec-23					
4	**Term & Reversion**							
6		Start Date	Valuation Term	Slice Type	Yield	Deferred	Net Rent	YP
7		31/12/2023	4.6167 yrs	Term	6.00%	0 yrs 0 mths	215,000	
8		12/08/2028	In Perp Def	Reversion	8.00%	4 yrs 7 mths	273,690	

Figure 3.10 Term & Reversion Spreadsheet 6.

YP (Years Purchase) will have the formulas as defined before.
For the term:
= ((1+yield)^-term)/yield
For the reversion:
= 1/yield
But then further PV'ed as the perpetuity is deferred.

B	C	D	E	F	G	H	I
	Valuation Date:	31-Dec-23					
	Term & Reversion						
	Start Date	Valuation Term	Slice Type	Yield	Deferred	Net Rent	YP
	31/12/2023	4.6167 yrs	Term	6.00%	0 yrs 0 mths	215,000	=(1-(1+F7)^-D7)/F7
	12/08/2028	In Perp Def	Reversion	8.00%	4 yrs 7 mths	273,690	=1/F8

Figure 3.11 Term & Reversion Spreadsheet 7.

CF Def is the deferred period (same as in Column G) but shown in years. The reason for this is because we added text in the deferred formula (Column G), consequently, this cell cannot be used in a formula, so we need to convert it into number or fraction of years in order to calculate the PV factor.

Again, **Deferred** is the distance between the valuation date and the start of the applicable rent. We will use the YEARFRAC as in the **Valuation Term** (Column D).

B	C	D	E	F	G	H	I	J
	Valuation Date:	31-Dec-23						
	Term & Reversion							
	Start Date	Valuation Term	Slice Type	Yield	Deferred	Net Rent	YP	CF Def.
	31/12/2023	4.6167 yrs	Term	6.00%	0 yrs 0 mths	215,000	3.93	=YEARFRAC($E$2,C7)
	12/08/2028	In Perp Def	Reversion	8.00%	4 yrs 7 mths	273,690	12.50	=YEARFRAC($E$2,C8)

Figure 3.12 Term & Reversion Spreadsheet 8.

PV is the present value factor that should be applied to the **YP** in case the rent is deferred. The formula of PV is:

$$PV = (1 + \text{yield})^{-\text{def}}$$

B	C	D	E	F	G	H	I	J	K	L
	Valuation Date:	31-Dec-23								
	Term & Reversion									
	Start Date	Valuation Term	Slice Type	Yield	Deferred	Net Rent	YP	CF Def.	PV	Gross Value
	31/12/2023	4.6167 yrs	Term	6.00%	0 yrs 0 mths	215,000	3.93	0.00 yrs	=(1+F7)^-J7	
	12/08/2028	In Perp Def	Reversion	8.00%	4 yrs 7 mths	273,690	12.50	4.62 yrs	=(1+F8)^-J8	

Figure 3.13 Term & Reversion Spreadsheet 9.

Gross value is the capitalised value of the passing and market rents. And it is simply:

Gross Value = Net Rent × YP × PV

The reason why it is gross and not net is because we are assuming a net yield, but if the yields were gross, the value would be net. However, note that in this part of the book (up until the multi-let model), we will assume that the 'world is perfect', i.e., there are no transaction costs and whatever the buyer pays, the seller will receive, which means that both gross and net yields are the same.

Start Date	Valuation Term	Slice Type	Yield	Deferred	Net Rent	YP	CF Def.	PV	Gross Value
31/12/2023	4.6167 yrs	Term	6.00%	0 yrs 0 mths	215,000	3.93	0.00 yrs	1.0000	=H7*I7*K7
12/08/2028	In Perp Def	Reversion	8.00%	4 yrs 7 mths	273,690	12.50	4.62 yrs	0.7010	=H8*I8*K8

Valuation Date: 31-Dec-23

Term & Reversion

Figure 3.14 Term & Reversion Spreadsheet 10.

Valuation Date: 31-Dec-23

Term & Reversion

Start Date	Valuation Term	Slice Type	Yield	Deferred	Net Rent	YP	CF Def.	PV	Gross Value
31/12/2023	4.6167 yrs	Term	6.00%	0 yrs 0 mths	215,000	3.93	0.00 yrs	1.0000	845,176
12/08/2028	In Perp Def	Reversion	8.00%	4 yrs 7 mths	273,690	12.50	4.62 yrs	0.7010	2,398,074
									3,243,250

Figure 3.15 Term & Reversion Spreadsheet 11.

Best practice tips:

Rename the sheet Term & Rev
Format and colour code (explained further in the chapter)
Protect the sheet

Don't forget to save the work if AutoSave isn't on.

Real Estate Valuation Using Capitalisation Methods

Hardcore Method

EXERCISE 2

Using the same Case Study 1, value the property using the Hardcore method.

If you are creating the model along with the book, you should now be able to copy the Term & Reversion sheet and rename it Hardcore as the structure is the same. The only difference is that we will need to add an extra column in **I** for the **Froth.**

> **Add column**
>
> **Froth** = Market Rent – Passing Rent

Start Date	Valuation Term	Slice Type	Yield	Deferred	Net Rent	Froth	YP	CF Def.	PV	Gross Value
31/12/2023		Hardcore	6.00%	0 yrs 0 mths	215,000			0.00 yrs	1.0000	0
12/08/2028	In Perp Def	Reversion (Top Layer)	8.00%	4 yrs 7 mths	273,690		12.50	4.62 yrs	0.7010	0
										0

Figure 3.16 Hardcore Spreadsheet 1.

For the Hardcore method, the valuation term is now In Perp and the YP is the day 1 perpetuity, = 1/yield. Whereas the Froth is the Market Rent – Passing Rent

Start Date	Valuation Term	Slice Type	Yield	Deferred	Net Rent	Froth	YP	CF Def.	PV	Gross Value
31/12/2023	In Perp	Hardcore	6.00%	0 yrs 0 mths	215,000		=1/F7	0.00 yrs	1.0000	3,583,333
12/08/2028	In Perp Def	Reversion (Top Layer)	8.00%	4 yrs 7 mths	273,690	=H8-H7	12.50	4.62 yrs	0.7010	514,242
										4,097,576

Figure 3.17 Hardcore Spreadsheet 2.

	Start Date	Valuation Term	Slice Type	Yield	Deferred	Net Rent	Froth	YP	CF Def.	PV	Gross Value
	31/12/2023	In Perp	Hardcore	6.00%	0 yrs 0 mths	215,000		16.67	0.00 yrs	1.0000	3,583,333
	12/08/2028	In Perp Def	Reversion (Top Layer)	8.00%	4 yrs 7 mths	273,690	58,690	12.50	4.62 yrs	0.7010	514,242
											4,097,576

Valuation Date: 31-Dec-23

Figure 3.18 Hardcore Spreadsheet 3.

Term and Reversion vs Hardcore: Which One to Use?

In terms of valuation, especially if doing a so-called 'Red Book' Valuation per RICS (Royal Institution of Chartered Surveyors) terms, the valuer is free to use their discretion in choosing the method of valuation they think is most appropriate.

In our case study, both methods yielded very different values. The Term & Reversion gave a value of £3.2 million, whereas the Hardcore gave a higher value of £4.1 million. The mathematical reason for the difference is because the lower initial yield is applied to a greater portion of the income in the Hardcore method, therefore resulting in a higher value. But the question is: which one to use?

My joke with my students is that you should use the Term & Reversion when you are buying and the Hardcore method when you are selling; but it is a joke as you see because this lacks scientific explanation. The way I rationalise the use of each method is by looking at the lease events that will give rise to the property reversion. If the lease event is a break or expiry, then I should use the Term & Reversion method, because the different yields will reflect both the market and tenants' risks better, since the covenants are unknown for the new tenant, and any voids, rent-free, or capex are implied in the higher reversionary yield.

On the other hand, if the lease event is a review to market, especially if upwards only, or an assumed renewal, then the Hardcore method seems more appropriate, and a higher value is also justified given that there is less risk in the income stream, post lease event.

Over-Rented

An over-rented scenario occurs when the passing rent is higher than the market rent. This happens in recessive scenarios, when there is more supply of space than demand for it, or when there has been a continued development boom, and competition for tenants has increased.

Term & Reversion

In situations when passing rent is higher than market rents, there is an increased risk of break or the tenant possibly renegotiating the lease downwards. In this scenario, we need to adjust the initial yield upwards in order to reflect the increased risk.

On the other hand, the reversionary yield should be adjusted downwards to reflect the lower ERV.

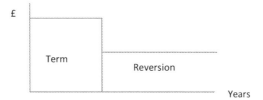

Figure 3.19 Term And Reversion Illustration.

Layer Method

Figure 3.20 Layer Method Illustration.

Valuers in practice started to approach the challenge of over-rented properties by adapting the layer method and 'top-slicing' the portion of the contractual rent they considered to be in excess of the current rental value. They capitalised the core income as if the property was fully let at the appropriate capitalisation rate. They then capitalised the top-slice income for the unexpired term of the lease at a rate that reflected the fact that it was a fixed income and that it was dependent on the tenant's ability to continue to pay the rent.

Years Purchase for Over-Rented Scenarios

	YP	Formula
Term	YP for n years	$=(1-(1+yield)^{-years})\,/\,yield$
Reversion	YP in perp def n years	$=(1+yield)^{-years}\,/\,yield$ *or* PV of $1/yield$
Bottom Layer	YP in perp	$=1/yield$

Figure 3.21 Years Purchase Calculations.

Where *years* is the number of years to lease expiry and *yield* is the appropriate property yield.

Term & Reversion

CASE STUDY 2

Retail Single-Let West End Investment, London W1,

Investment Summary

- Long leasehold of over 130 years
- Prime Oxford Street retail pitch next to Bond Street Station
- Major shoe retailer flagship store in the West End of London
- A key beneficiary of the recently opened Elizabeth Line (Crossrail)
- The tenant has recently refitted the store to the highest of standards
- Total area of 5,449 sq ft (506 sq m) over ground, lower ground and part first floors
- Passing rent of £1,600,000 per annum equating to a passing rent of £992 Zone A
- 100% let to Shoe retailer – one of the most exclusive shoe retailers on the high street
- Long-term income security with a lease until January 2038 (c. 15 years unexpired)
- The next rent review is due on 11 January 2028

The valuation date is 31 December 2023 and the contract will expire on 10 January 2038. The ERV is £925 Zone A, equating to an ERV of £1,495,000. The valuer has estimated a comparable equivalent yield of 5.0% and believes that the initial yield on the passing rent (over-rented) should be 6.0 given that the tenant may renegotiate a lower rent.

Exercise 3

Value this property using the Term & Reversion method.

For the Term & Reversion, you can use the same calculation as for the under-rented scenario.

Start Date	Valuation Term	Slice Type	Yield	Deferred	Net Rent	YP	CF Def.	PV	Gross Value
31/12/2023	4.0306 yrs	Term	6.00%	0 yrs 0 mths	1,600,000	3.49	0.00 yrs	1.0000	5,581,743
11/01/2028	In Perp Def	Reversion	5.00%	4 yrs 0 mths	1,495,000	20.00	4.03 yrs	0.8215	24,562,159
									30,143,902

Figure 3.22 Term and Reversion Spreadsheet 12.

Valuation Date: 45291

Term & Reversion

Start Date	Valuation Term	Slice Type	Yield	Deferred	Net Rent	YP	CF Def.	PV	Gross Value
45291	=YEARFRAC(C7,C8)	Term	0.06	=DATEDIF($E$2,C7,"y")&" yrs "&DATEDIF($E$2,C7,"ym")&" mths"	1600000	=(1-(1+F7)^-D7)/F7	=YEARFRAC($E$2,C7)	=(1+F7)^-J7	=H7*I7*K7
46763	In Perp Def	Reversion	0.05	=DATEDIF($E$2,C8,"y")&" yrs "&DATEDIF($E$2,C8,"ym")&" mths"	1495000	=1/F8	=YEARFRAC($E$2,C8)	=(1+F8)^-J8	=H8*I8*K8
									=SUM(L7:L8)

Figure 3.23 Term and Reversion Spreadsheet 13.

Hardcore Method

EXERCISE 4

Following the Case Study 2 assumptions, value the property using the Hardcore method.

Start Date	Valuation Term	Slice Type	Yield	Deferred	Net Rent	Froth	YP	CF Def.	PV	Gross Value
31/12/2023	In Perp	Hardcore	5.00%	0 yrs 0 mths	1,495,000		20.00	0.00 yrs	1.0000	29,900,000
31/12/2023	5.0000 yrs	Froth	6.00%	0 yrs 0 mths	1,600,000	105,000	4.21	0.00 yrs	1.0000	442,298
										30,342,298

Valuation Date: 45291   Reversion Date: 47118

Hardcore

Start Date	Valuation Term	Slice Type	Yield	Deferred	Net Rent	Froth	YP	CF Def.	PV	Gross Value
45291	In Perp	Hardcore	0.05	=DATEDIF($E$2,C7,"y")&" yrs "&DATEDIF($E$2,C7,"ym")&" mths"	1495000		=1/F7	=YEARFRAC(E2,C7)	=(1+F7)^-K7	=H7*J7*L7
=C7	=YEARFRAC(C8,I2)	Froth	0.06	=DATEDIF($E$2,C8,"y")&" yrs "&DATEDIF($E$2,C8,"ym")&" mths"	1600000	=H8-H7	=(1-(1+F8)^-D8)/F8	=YEARFRAC(E2,C8)	=(1+F8)^-K8	=I8*J8*L8
										=SUM(M7:M8)

Figure 3.24 Hardcore Spreadsheet 4.

Rack Rented

When the passing rent is the same as the market rent, either because the market went down and up again, up and down again or simply because there has just been a rent review to ERV or it is a new lease, we refer to this property being rack rented (at least in the United Kingdom).

All-Risks Yield Method or True Capitalisation Method

In fact, this is just like the hardcore method but without the top layer:

Figure 3.25 Rack-Rented Illustration.

In this case, we can use a straight capitalisation method:

	YP	Formula	Yield
Bottom Layer	YP in perpetuity	=1/yield	Straight comparison with market initial yields

Figure 3.26 Straight Capitalisation Method.

This method can also be applied when we lack market evidence that the property is under- or over-rented. Then, the equivalent yield is applied to just the initial income stream or passing rent.

EXERCISE 5

A freehold office of good quality is let at a full repairing and insuring (FRI) rent of 80,000 p.a. There are 5 years left for the expiry of the lease contract, and the ERV is the same 80,000. An identical property has just been sold next door at a 7% equivalent yield. Value this property using the All-Risks Yield method.

Start Date	Valuation Term	Slice Type	Yield	Deferred	Net Rent	Froth	YP	CF Def.	PV	Gross Value
31/12/2023	In Perp	Hardcore	7.00%	0 yrs 0 mths	80,000		14.29	0.00 yrs	1.0000	1,142,857
										1,142,857

Valuation Date: 31-Dec-23

Figure 3.27 All-Risks Yield Model 1.

Valuation Date: 45291

Start Date	Valuation Term	Slice Type	Yield	Deferred	Net Rent	Froth	YP	CF Def.	PV	Gross Value
45291	In Perp	Hardcore	0.07	=DATEDIF($E$2,C7,"y")&" yrs "&DATEDIF($E$2,C7,"ym")&" mths"	80000		=1/F7	=YEARFRAC(E2,C7)	=(1+F7)^-K7	=H7*J7*L7
										=M7

Figure 3.28 All-Risks Yield Model 2.

Output Yields

You may have realised from this section that yields in the property market are plenty and confusing. But knowing which yield is used and when will demystify the jargon related to yields: initial, reversionary, exit, and equivalent.

Now that you know that the value of the property is based on input yields, we can calculate output yields, namely the initial, reversionary, and equivalent yields.

In order to calculate the equivalent yield in Excel, we will need to solve both yields: the term and the reversionary yield, given the property valuation we obtained above.

We will use the Case Study 1 (17 Fleet Street), Exercise 1 (Term & Reversion) to calculate the output yields. I am copying the valuation calculations below for ease of reference:

	B	C	D	E	F	G	H	I	J	K	L
1											
2		Valuation Date:		31-Dec-23							
3											
4		**Term & Reversion**									
5											
6		Start Date	Valuation Term	Slice Type	Yield	Deferred	Net Rent	YP	CF Def.	PV	Gross Value
7		31/12/2023	4.6167 yrs	Term	6.00%	0 yrs 0 mths	215,000	3.93	0.00 yrs	1.0000	845,176
8		12/08/2028	In Perp Def	Reversion	8.00%	4 yrs 7 mths	273,690	12.50	4.62 yrs	0.7010	2,398,074
9											3,243,250
10											

Figure 3.29 Term and Reversion Model 1.

Initial Yield

Passing Rent/Value
  = 215,000/3,243,250 = 6.63%

Equivalent Yield

There are two ways of calculating the equivalent yield.

Version one, which is the one used by Argus Enterprise (a real estate modelling software), that is the average of the initial and reversionary yields weighted by their respective values. In our case, it would be:

  = (initial yield × term value + reversionary yield × reversion value)/total value
  = (6.0% × 845,176 + 8.0% × 2,398,074)/3,243,250 = 7.478%

In Excel, we will use the SUMPRODUCT function:

  = SUMPRODUCT((yields range) × (values range))/total value

What the SUMPRODUCT does is that it calculates a matrix, multiplying the corresponding values and adding them up at the end. So:

Yields	Values	Multiply
6.0%	845,146	6.0% x 845,176 = 50,711
8.0%	2,398,074	8.0% x 2,398,074 = 191,846
	SUMPRODUCT answer:	242,556

Figure 3.30 Equivalent Table 1.

By dividing the SUMPRODUCT answer by the total value gives us the equivalent yield as calculated by Argus Enterprise.[2]

Equivalent Yield = 242,556/3,243,250 = 7.478%

However, this equivalent yield does not provide us with the right answer since if you use the same yield as initial and reversionary, the value will not be the same as if calculated with the input yields of 6.0 and 8.0%. See below:

Start Date	Valuation Term	Slice Type	Yield	Deferred	Net Rent	YP	CF Def.	PV	Gross Value
31/12/2023	4.6167 yrs	Term	7.48%	0 yrs 0 mths	215,000	3.79	0.00 yrs	1.0000	814,165
12/08/2028	In Perp Def	Reversion	7.48%	4 yrs 7 mths	273,690	13.37	4.62 yrs	0.7168	2,623,127
									3,437,292

Valuation Date: 31-Dec-23

Figure 3.31 Equivalent Model 1.

In order to achieve the 'true' equivalent yield, we need to solve for making both initial and reversionary yields the same and targeting the original value of 3,243,250.

For this, I will show how to use the **Goal Seek** function in Excel.

Steps

1 Copy the Term & Reversion sheet so you have a separate sheet for the equivalent yield calculation only.
2 In the new sheet, copy and paste as value the target total value (3,243,250).
3 Make the reversionary yield equals to the initial yield.

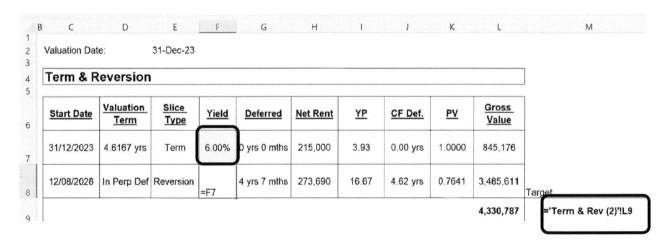

Figure 3.32 Equivalent Model 2.

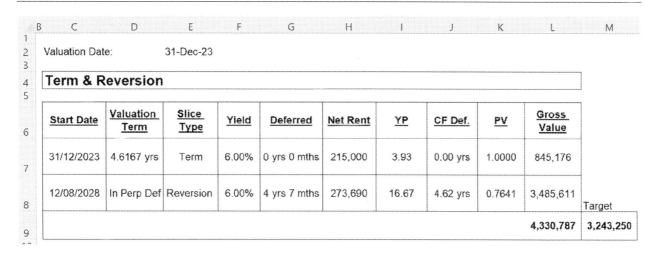

Figure 3.33 Equivalent Model 3.

4 Go to Data > What-If Analysis > Goal Seek:

Figure 3.34 Equivalent Model 4.

5 **Set Cell:** total value (cell L9)
6 **To Value:** 3243250 (the target value as calculated by the input yields)
7 **By changing cell:** Initial Yield (cell F7)
8 Click **OK**

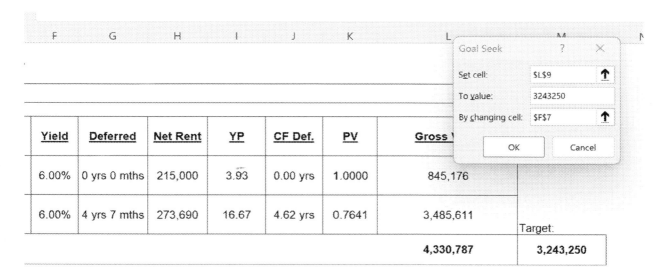

Figure 3.35 Equivalent Model 5.

34 Real Estate Financial Modelling in Excel

The **Yields** will change to 7.90% in both cells F7 and F8. This is the 'true' equivalent yield.

	Start Date	Valuation Term	Slice Type	Yield	Deferred	Net Rent	YP	CF Def.	PV	Gross Value	
Valuation Date:	31-Dec-23										
Term & Reversion											
	31/12/2023	4.6167 yrs	Term	7.90%	0 yrs 0 mths	215,000	3.75	0.00 yrs	1.0000	805,607	
	12/08/2028	In Perp Def	Reversion	7.90%	4 yrs 7 mths	273,690	12.65	4.62 yrs	0.7039	2,437,643	Target
										3,243,250	3,243,250

Figure 3.36 Equivalent Model 6.

Yield Analysis and Net Value

Now that we know how to calculate the output yields, we can create a comparative table with all relevant yields in our valuation.

The final report could look something like this:

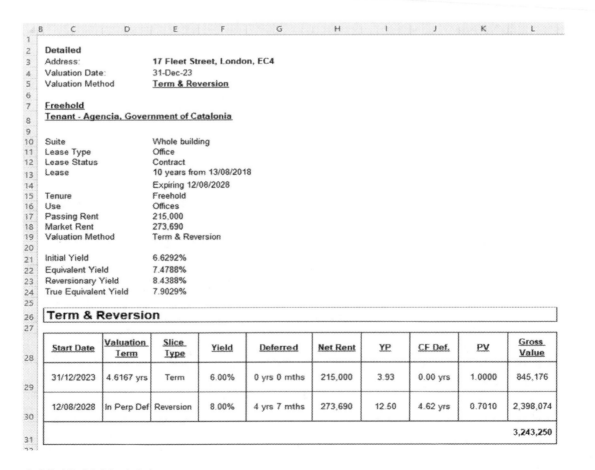

Figure 3.37 Yield Model 1.

Real Estate Valuation Using Capitalisation Methods

What Can You Say about the Property Value Looking at the Yields Alone?

In this example, which the initial yield is 6.63%, and the equivalent yield is 7.90%, the initial yield is therefore lower than the equivalent yield. This tells us that the property is **under-rented and there is potential for rent uplifts**.

If You Assume Purchaser's Costs of 7.0%, What Would Be the Net Value of the Asset?

For you to calculate the net value of the property, considering the yields are all quoted on a net basis, then you need to divide the gross value by (1 + purchaser's costs).
Net Value = Gross Value/(1 + purchaser's costs)
Net Value = 3,243,250/(1 + 7%) = 3,031,074

Formatting

The next step is formatting the cells so that we can easily differentiate the input from the output cells. The input cells being those that the user can change and the output cells being those that contain formulas and that typically should not be changed by the user.

In presentations of our models so far, our sheets have already been formatted accordingly (so that this manual would look nice). But when we construct models, we tend to leave the formatting for the end.

EXERCISE 6

Using an unformatted version of Exercise 1 calculations, format the worksheet appropriately.
Starting from the unformatted Exercise 1:

	B	C	D	E	F	G	H	I	J	K	L
1											
2		Valuation Date:		45291							
3											
4		Term & Reversion									
5											
6		Start Date	Valuation Term	Slice Type	Yield	Deferred	Net Rent	YP	CF Def.	PV	Gross Value
7		45291	5	Term	0	0 yrs 0 mths	215000	4	0	1	845176
8		46977	In Perp Def	Reversion	0	4 yrs 7 mths	273690	13	5	1	2398074
9											3243250
10											

Figure 3.38 Formatting 1.

We can then merge cells for the title (Term & Reversion), change background, font colour, and font size. As it is basic Excel knowledge, I will assume you know how to do it and choose colours, bold, and sizes according to your taste. I am more concerned about the numbers formatting in this course.

Dates: Format a Number as Dates

Excel stores dates as numbers; to change the formatting, just press **CTRL + #**

	B	C	D	E	F	G	H	I	J	K	L
1											
2		Valuation Date:		31-Dec-23							
3											
4		Term & Reversion									
5											
6		Start Date	Valuation Term	Slice Type	Yield	Deferred	Net Rent	YP	CF Def.	PV	Gross Value
7		31-Dec-23	5	Term	0	0 yrs 0 mths	215000	4	0	1	845176
8		12-Aug-28	In Perp Def	Reversion	0	4 yrs 7 mths	273690	13	5	1	2398074
9											3243250

Figure 3.39 Formatting 2.

Valuation Term: Format a Number to Include a Text Next to It

If you just add **yrs** next to the number, you will have a problem because Excel will not recognise this cell as a number anymore and your calculation will not work. We need to keep 5 as a number but add the text next to it. How?

- In the relevant cell, click **CTRL + 1** (the **Format Cells** window will display)
- Input the variables as per window below

Check that you are in the **Number** tab, **Custom** and your **Type** field has the same text as below. Note that the sign before and after **yrs** is the double speech mark, or **Shift + 2**

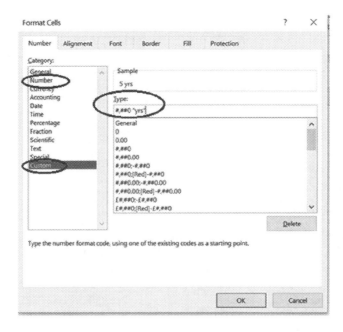

Figure 3.40 Format Cells - Number.

Real Estate Valuation Using Capitalisation Methods 37

Yield: Format Percentages:

For formatting percentages, you can use the ribbon % sign:

- Go to **Home** > **%**
- Then, you can add decimal places (2), by clicking twice.

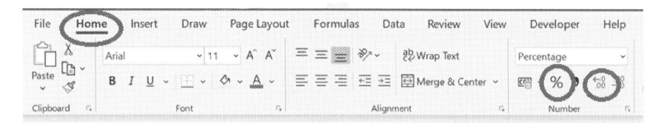

Figure 3.41 Format Percentages.

Or you can use shortcut:

- For percentage: **CTRL + Shift + 5**
- For decimals: **CTRL + 1** >>> Tab Tab **2** > **ENTER**

Passing Rent, YP and Values: Format with Thousands Separator and Adjust Decimal Places

In the cell where you want to add thousands separator, so that number instead of looking 1000 will look like 1,000; just press

- **CTRL + Shift + 1.**

Then, the number will now look like 1,000.00. It means that it also adds two (2) decimal places.
If you want to get rid of the decimal places, just use the decimals command again but change to zero (0) the number of decimal places you want.

- For decimals: **CTRL + 1** >>> Tab Tab **Number of decimals** > **ENTER**

Borders: Add Outside Borders

In financial models, as we like to see numbers in blocks, for example, inputs in a block, outputs in another block; Term calculations as one block, Reversion as another, etc.; outside borders are good visual delimiters to separate these blocks.

You just need to highlight the area you want to be delimited by the borders and press **CTRL + Shift + 7**.

	Valuation Date:		31-Dec-23							
Term & Reversion										
Start Date	Valuation Term	Slice Type	Yield	Deferred	Net Rent	YP	CF Def.	PV	Gross Value
31-Dec-23	4.6167 yrs	Term	6.00%	0 yrs 0 mths	215,000	3.93	0.00 yrs	1.0000	845,176
12-Aug-28	In Perp Def	Reversion	8.00%	4 yrs 7 mths	273,690	12.50	4.62 yrs	0.7010	2,398,074
									3,243,250

Figure 3.42 Format Borders.

Changing Input and Output Cells Colours

We are getting there. We just need to change the input cell colours so that our spreadsheet is transparent and user-friendly, as per best practice.

As this is a small worksheet, you could be tempted to go cell by cell looking for input cells, right? I would. But that's absolutely not the point; instead, we want to find *efficient* ways to deal with Excel (imagine you had a worksheet with 300 input cells. Would you go one by one? The answer is NO!!). Instead:

- Click **F5**

A window will display:

Figure 3.43 Go To.

- Go to **Special**

Next window:

Figure 3.44 Go To Special.

- Tick **Constants**
- Untick **Text**, **Logicals**, and **Errors**
- Click **OK**

Real Estate Valuation Using Capitalisation Methods 39

Highlighted cells will appear:

Start Date	Valuation Term	Slice Type	Yield	Deferred	Net Rent	YP	CF Def.	PV	Gross Value
31-Dec-23	4.6167 yrs	Term	6.00%	0 yrs 0 mths	215,000	3.93	0.00 yrs	1.0000	845,176
12-Aug-28	In Perp Def	Reversion	8.00%	4 yrs 7 mths	273,690	12.50	4.62 yrs	0.7010	2,398,074
									3,243,250

Valuation Date: 31-Dec-23

Term & Reversion

Figure 3.45 Constants.

These are the input cells.

With the cells still highlighted (i.e., DO NOT click anywhere else on the sheet), go to:

Home > **Cell Styles** > choose your format

I select 'Input' (orange background with blue font) because this is what Excel is already telling us that is how an 'Input' is, so I will not fight against. But if you think you would like to add another format that is less bright, then you can choose the cell style accordingly or even add a **New Cell Style** according to your preference.

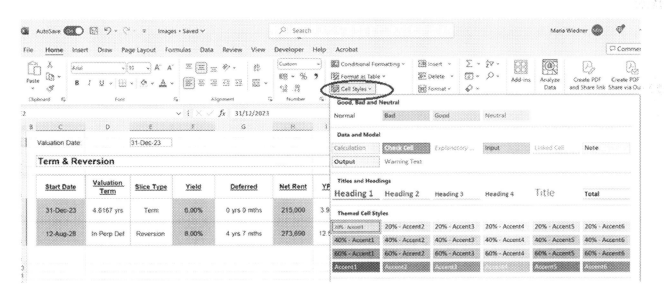

Figure 3.46 Cell Styles.

Your table will now look like this:

Valuation Date: 31-Dec-23

Term & Reversion

Start Date	Valuation Term	Slice Type	Yield	Deferred	Net Rent	YP	CF Def.	PV	Gross Value
31-Dec-23	4.6167 yrs	Term	6.00%	0 yrs 0 mths	215,000	3.93	0.00 yrs	1.0000	845,176
12-Aug-28	In Perp Def	Reversion	8.00%	4 yrs 7 mths	273,690	12.50	4.62 yrs	0.7010	2,398,074
									3,243,250

Figure 3.47 Cell Styles Applied.

With the input cells still ACTIVE, i.e., they are still being selected (you will see them being shaded):

- **Hover** over any of the active cells and **right-click**
- Go to **Format Cells** from the drop-down menu
- Click on tab **Protection**
- Untick **Locked**
- Click **OK**

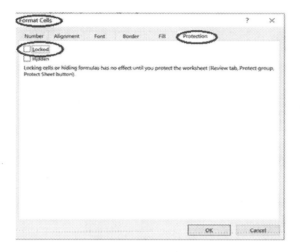

Figure 3.48 Cell Styles Protected.

- Click on any cell in the sheet

The steps above unlocked the input cells, so that when we protect the worksheet, we are still able to change the inputs even though the sheet is locked.

Hiding the Gridlines

To improve how the financial model looks, it is common practice to remove gridlines at the end of the process. This is, of course, a matter of taste, but at least you know how to do it.

- Go to **View**
- Untick Gridlines

Figure 3.49 Hide Gridlines.

Voila:

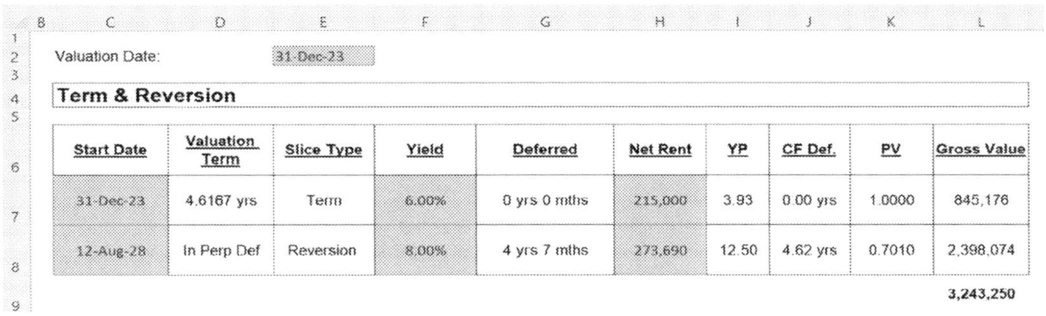

Figure 3.50 Hide Gridlines 2.

Protecting Sheet

The only missing step is to protect the worksheet so that your users don't mess around with the formulas.

Now that you have unlocked the input cells, you can lock the worksheet.
- Go to Review > Protect Sheet

Figure 3.51 Lock Sheet.

- Enter the same password twice

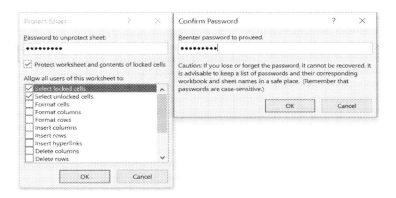

Figure 3.52 Lock Sheet 2.

If you want to test if the protection is working, you can try to change the data of any input cells (the highlighted cells). You should be able to change these. But if you want to change the values of any output cells (non-highlighted cells), you should get the error message below:

42 Real Estate Financial Modelling in Excel

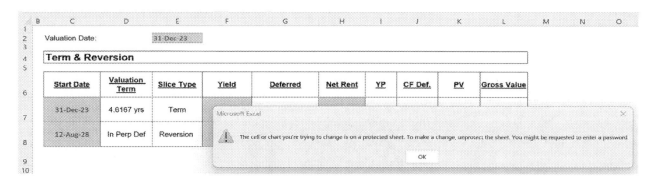

Figure 3.53 Lock Sheet 3.

Passwords: make sure you have a log of passwords. There are some ways to unlock sheets without knowing the passwords, but it's not the most ethical way to deal with spreadsheets.

Note that you will need to protect EACH sheet in the workbook as this cannot be done 'in bulk', i.e., all at the same time.

Also, a note on **Protect Workbook**: this only protects the STRUCTURE of the workbook and not cells in worksheets. By using this facility, you won't be able to change names of sheets, hide/unhide them, and add/delete sheets. Everything else within sheets is possible, hence the need to **Protect Worksheets** individually.

Conclusion

This chapter showed you the building blocks to value properties in under-, over- and rack-rented properties using the capitalisation method for property valuations. We demonstrated how valuers use the Term & Reversion, Hardcore, and Rack-Rented methods to ascertain property values based on comparable evidence.

The main rationale for using implicit methods is that they are simple and easy to explain and that yields derive from comparable transaction evidence in the market. We further showed how to calculate output yields, such as initial and equivalent yields.

This chapter also explained how to format models with the ribbons and shortcuts, and how to protect your sheet so inputs values can be changed, but formulas for output calculations can be protected.

Notes

1 Discounted Cash Flow Valuations, RICS, November 2023.
2 Software for property valuation and asset management. https://argus1.altusgroup.com/argus-enterprise-demo/.

Chapter 4

Introduction to Discounted Cash Flow Modelling and Analysis

Chapter Contents

Introduction	43
Single-Let Cash Flows	44
Annual Cash Flow	44
Fixed Rent over the Holding Period	44
Layout	44
Inputs	45
Outputs	49
Output Table	50
Complete Model	50
How to Reconcile the Implicit Valuation with the DCF?	51
Quarterly Discounted Cash Flow	52
Single Rent Review or Renewal to ERV	52
Layout	52
Inputs	53
Cash Flow	54
Outputs	55
Review Cycles	56
Layout	56
Inputs	57
Outputs	58
Indexation to Inflation	59
Layout	59
Inputs	60
Cash Flow	60
Output Table	61
Is It a Good Investment?	62
Conclusion	62

Introduction

Discounted cash flow (DCF) is a well-known investment tool used to determine the internal rate of return (IRR) of an investment, net present value (NPV), and many other financial measures. It is used as a market and investment valuation tool in many markets where the valuer can support all variables used by reference to relevant market data and their own assumptions. Investors can use a DCF model to make investment decisions to buy, sell, or hold a certain asset and find the 'fair value'.

In real estate, DCF is the most used tool in investment and underwriting processes, in which the investor needs to investigate the worth and investment case of an asset.

The Royal Institution of Chartered Surveyors (RICS) is also encouraging valuers to adopt the DCF for real estate investment valuation purposes, and as such, it is crucial that surveyors now become more adept to do a DCF valuation for 'Red Book' purposes as well.

We will start modelling a DCF for a single-let property first and link it to the conventional valuation methods so we can compare and contrast.

Single-Let Cash Flows

Single-let cash flows will take place when a property is let to a single tenant. This can be a single-family residential property, a high street shop, an office block with just one tenant, a warehouse, or anything else you may come across with just one tenant in place.

The importance of differentiating single from multi-tenanted properties is that in single-let properties, all inputs are 'global' inputs, i.e., the assumptions will be the same throughout the cash flow, whereas with multi-tenanted properties, we will need to distinguish between 'global' inputs and 'tenant' inputs (the tenancy schedule).

We will first start with an annual cash flow with no lease events as such and build it up step by step incorporating different lease events that are relevant to the real estate investment valuation using a DCF.

Annual Cash Flow

We will replicate Case Study 1 here (17 Fleet Street) so we can then compare with the Implicit Valuation methods.

EXERCISE 7

Fixed Rent over the Holding Period

Passing rent £215,000 p.a. with 5-yearly review to open market rents and estimated rental value (ERV) is £273,690. Assume for now that this rent is paid annually in arrears, and in line with our assumptions in the Term and Reversion, the initial yield was 6.63%. The investor estimates that the average rental growth will be 2.00% p.a. for the next 5 years, and the exit yield will be 200 basis points greater to account for the need of refurbishment in the building and shorter lease length.

The investor's target rate is 8.00% for this property, assuming the costs of money and risks for this investment.

Assume a holding period of five years and no other lease events or capital expenditures.

Layout

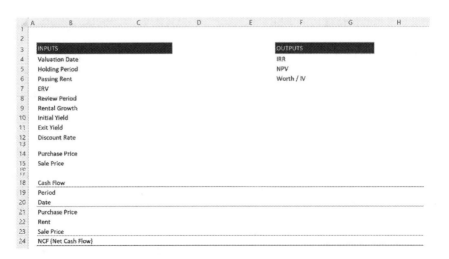

Figure 4.1 Cash Flow 1.

Inputs

1 **Holding period**
2 This is an investment horizon, meaning that we will buy this property at period 0 (today) and sell it at the end of the holding period. We will assume a 5 (five) year holding period.
3 To format as a number but showing 'yrs' after the number:

Right-click on the cell
Go to **Format Cells**
Click on tab **Number**
Click on **Custom**
In **Type** field, type **#,##0 "yrs"**

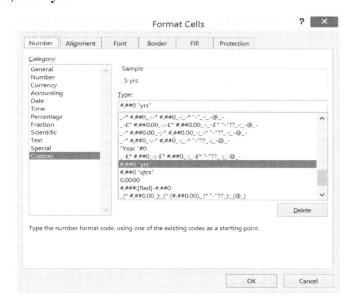

Figure 4.2 Cash Flow 2.

4 **Passing rent**
5 This is the rent being received at the time of purchase
6 You will need to type the passing rent, i.e. 215000

To format it with comma and decimal places, click **CTRL + Shift + 1**
This will now be formatted as 215,000.00
If you want, you can take out the two decimal places by then typing **CTRL +1**
Tab > Tab > 0 (zero)

7 **ERV/market rent**
8 The estimated rental value is, in this case, the same as the passing rent because the property has a new lease in place. You can type the value of passing rent again of 273,690 as hardcoded number and format in the same way as the passing rent.
9 **Review period**
10 This is the input for when the review period will take place. In this case, with no review, we will assume it is at the end of the holding period, or year 5.
11 Type in 5 and format as per the holding period with "**yrs**".
12 **Rental growth**
13 Assume rental growth of 2.00%.

14 Type in 0.02 and then:

Press **CTRL + Shift + 5**
CTRL + 1
Tab > Tab > 2

15 **Initial yield**
16 Enter the relevant initial yield as per comparable in the market.
17 **Exit yield**
18 Enter the estimated exit yield at the time of sale of the investment.
19 **Discount rate**
20 The discount rate is estimated at 8.00% for this type of investment.
21 **Purchase price**
22 = passing rent/initial yield
23 **Sale price**
24 The sale price is calculated in the same way as the purchase price: rent/yield. However, we now need to solve for the rent at exit, which should consider rental growth in the market and the yield is based on the expectation of future property yields in the market.

Rent at Exit = ERV × (1 + growth)^period
To find the sale price, all that needs doing is capitalising the rent at exit by the yield, or
= ERV × (1 + growth) ^ (hold period)/exit yield

25 The **Inputs** table will now look like:

	A	B	C
3		INPUTS	
4		Valuation Date	31-Dec-23
5		Holding Period	5 yrs
6		Passing Rent	215,000.00
7		ERV	273,690.00
8		Review Period	5 yrs
9		Rental Growth	2.00%
10		Initial Yield	6.63%
11		Exit Yield	8.83%
12		Discount Rate	8.00%
13			
14		Purchase Price	3,242,836
15		Sale Price	3,422,150

Figure 4.3 Cash Flow 3.

26 **Cash flow**
27 We can now start creating the cash flow for this investment.
28 Remember that as per best practice, we will use time horizontally and cash flow labels vertically.
29 **Period**

- We will need to enter a series of periods, in this case from 0 to 5, and we will use the **Fill** function.

For the first period, type 0
Go back to the cell of the first period (where you will see 0)
Go to **Home > Fill > Series**

Introduction to Discounted Cash Flow Modelling and Analysis   47

Figure 4.4 Cash Flow 4.

Next window:

Figure 4.5 Cash Flow 5.

For the **Stop value**, type the holding period number, i.e. 5.

30 **Date**
31 For the date, put the valuation date.

To format as DD-MMM-YY, press **CTRL + #**

32 For date at time 1 to end of holding period, we will use the **EDATE** function:
33 The **EDATE** function stands for 'End Date' and will give you the date after the months you require. For instance, if you want to calculate the date 1 year after the start date, you will add 12 (because there are 12 months in any year), 3 for a quarter (the same logic applies), 1 for a month.
34 Therefore, the syntax for the **EDATE** function is:
=EDATE(start cell, months)
= EDATE (valuation date, period × 12)

Fix the valuation date using $$, so it is an absolute reference (see section **'Rent'** below for further information on fixing cells in Excel).

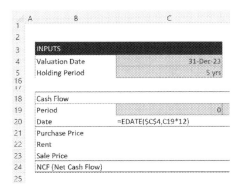

Figure 4.6 Cash Flow 6.

Copy the **EDATE** formula to the end of the holding period by highlighting the relevant cells and press **CTRL + R**

1. **Purchase price**
2. Just link the purchase price with its input cell.
3. BUT as this is money going out, we need to make it a negative value. Thus:
4. = - Purchase Price

5. **Rent**
6. As there is no review, meaning that the rents will be the same, all that is needed is linking the rent with its relevant cell.
7. Because the rent is received in arrears, we will first insert the rent in period 1. Also, as we will copy the formula to the right to show it in all relevant periods, we will need to 'fix' it, so the reference doesn't move. To do so, we will need to add the $ around the cell reference. This way, we ensure that as we copy the reference across, neither the column nor the row 'moves'.
8. =$Letter$Number
9. To insert the dollar sign automatically, all you need to do is to highlight the reference cell in the formula bar and press **F4 once**.

Figure 4.7 Fixing Cell Reference.

10. Remember:
11. This can only be done because in this exercise there are no rent reviews, so the rent will be unchanged throughout the holding period.
12. Adding dollar signs means that you will fix the reference cell so the formula will not 'travel' to the right when dragging the formula across.

13. **Sale price**
14. The sale price will be linked to the value we have created as an **Input.**
15. The only problem with the sale price, for the time being, is that we will need to add this value directly in period 5, when we exit the investment. Thus, in period 5, type in
16. = Sale Price
17. It is a positive number because we sell the investment, it is money coming in.

18. **Net cash flow (NCF)**
19. The NCF is simply the SUM of the cells above, from purchase price to sale price and passing by rent.
20. We can use the **Auto sum** function (from **Home** > **Editing** >∑) or the shortcut **ALT + =**

	A	B	C	D	E	F	G	H
17								
18		Cash Flow						
19		Period	0	1	2	3	4	5
20		Date	31-Dec-23	31-Dec-24	31-Dec-25	31-Dec-26	31-Dec-27	31-Dec-28
21		Purchase Price	-3,242,836					
22		Rent		215,000	215,000	215,000	215,000	215,000
23		Sale Price						3,422,150
24		NCF (Net Cash Flow)	=SUM(C21:C23)	215,000	215,000	215,000	3,637,150	
25								

Figure 4.8 Cash Flow 7.

21 Then, you can highlight all the NCF cells and press **CTRL + R**, or drag the formula to the right.
22 The complete cash flow will now be as follows:

	B	C	D	E	F	G	H
17							
18	Cash Flow						
19	Period	0	1	2	3	4	5
20	Date	31-Dec-23	31-Dec-24	31-Dec-25	31-Dec-26	31-Dec-27	31-Dec-28
21	Purchase Price	-3,242,836					
22	Rent		215,000	215,000	215,000	215,000	215,000
23	Sale Price						3,422,150
24	NCF (Net Cash	-3,242,836	215,000	215,000	215,000	215,000	3,637,150

Figure 4.9 Cash Flow 8.

	B	C	D	E
17				
18	Cash Flow			
19	Period	0	1	2
20	Date	=EDATE($C$4,C19*12)	=EDATE($C$4,D19*12)	=EDATE($C$4,E19*12)
21	Purchase Price	=-C14		
22	Rent		=$C$6	=$C$6
23	Sale Price			
24	NCF (Net Cash Flow)	=SUM(C21:C23)	=SUM(D21:D23)	=SUM(E21:E23)

Figure 4.10 Cash Flow 9.

23 This is your annual discounted cash flow model. Now, you need to add the Output: net present value (NPV), internal rate of return (IRR), and worth, which will then lead to the investment recommendation.

Outputs

To start our analysis, we will look at three values: IRR, NPV, and worth or investment value.

1 **IRR**
2 The internal rate of return is a metric which measures the profitability as a percentage of the potential investment. It should be looked on an annual basis, and as such, it gives you a yearly rate of return on investment.
3 The syntax in Excel is:
4 = IRR(values,[guess])
5 Values are your NCF row, and [Guess] is the value you think the IRR will be close to. The reason for [ ] is because this is not a required value and you can put 0.001 or your target rate. I tend to leave it out completely; however, the [Guess] value can be useful if Excel finds more than one IRR because either Excel will return an error value or will give you an IRR which may seem unrealistic.
6 The criticism about IRR is mainly that it will find more than one result at times; the assumption is that all cash flows can be reinvested at the same rate of IRR and assumes that investments can be equally replicated. For example, if you find a project of £1.00 returning an IRR of 100% and another project of £100, returning an IRR of 20%, in case you decide on IRR alone, you will choose the investment of £1, assuming it can be replicated 100 times.
7 Another point about the IRR is that results will always be shown based on the periodicity of the cash flow. If your cash flow is annual, the IRR will be shown as an annual rate; if your cash flow is quarterly, your IRR will be quarterly, and so on.

8 In our case in hand, it is not a problem because the cash flow is annual, but if this were not the case, we would need to compound it up to an annual figure.
9 **Net present value (NPV)**
10 The NPV is the sum of present values. The present values (PVs) are the discounted values of future cash flows, meaning that £100 in the future will worth LESS than today, so it's mainly the values taking into account the time value of the money (TVM).
11 The syntax in Excel is:
12 = NPV (discount rate, values)
13 You need to pay attention to two things:

Discount rate: this needs to be at the same periodicity as the cash flows. If you have built an annual cash flow, then the discount rate should be annual. If you have built a quarterly cash flow, you would need to de-compound the annual discount rate to a quarterly rate.

Values: when using the NPV function, Excel 'thinks' that all values happen at the end of the period, and the first value is discounted for one period already. The problem with this approach is that in our cash flow, the first value happens today, at time 0, so this should not be discounted. In this case, we need to exclude the first NCF from the value ranges and add it back after the NPV calculation.

14 =NPV (discount rate, NCF at period 1: NCF at period N) + Value at period 0
15 **Worth/investment value**
16 Worth is, in valuation jargon, the investment value to a particular investor. In that sense, it's the maximum price that this investor can pay for the asset, given all their assumptions entered into the cash flow calculation, such as exit yield, market growth, holding period, discount rate, etc.
17 It may be obvious by now that each investor will have their own set of assumptions, and as such, each investor will have different worth figures for the same asset.
18 The formula to be used to calculate the worth is:

Worth = NPV + Purchase Price

Output Table

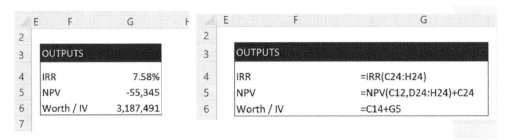

Figure 4.11 Cash Flow 10.

Complete Model

As per my own formatting guidelines, the completed model will look like the below. Note that the way I formatted the worksheet is not necessarily how you should do it, it is just what I think works for me, my taste, and abilities. You can change the colours, the distance between labels, and anything else. Just make sure it all looks pretty, simple, and elegant!

Introduction to Discounted Cash Flow Modelling and Analysis 51

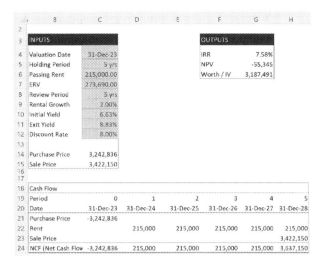

Figure 4.12 Cash Flow 11.

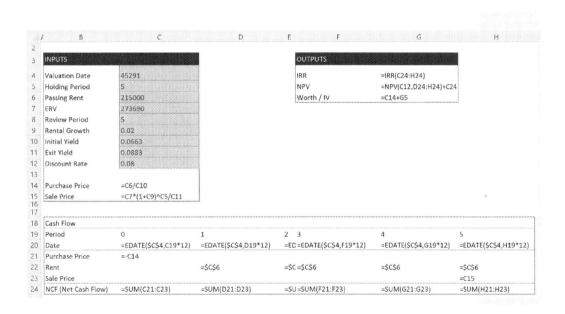

Figure 4.13 Cash Flow 12.

How to Reconcile the Implicit Valuation with the DCF?

In our case above, the DCF and the Term & Reversion method gave us different values: 3,187,491 and 3,242,836, respectively. If we consider the 'say' (rounded value) to be 3.2 million, then both gave us the same answer. However, one of the questions that valuers now have is how to find an appropriate discounted rate when you are used to the concept of yields/cap rates? The easiest way is to back-solve the value achieved by the implicit valuation by targeting the discount rate using **Goal Seek**. In this case, you would:

Data > What-If Analysis > Goal Seek
Set Call: Worth/IV
To: Value achieved via Implicit Method
By changing cell: Discount Rate

In our case:

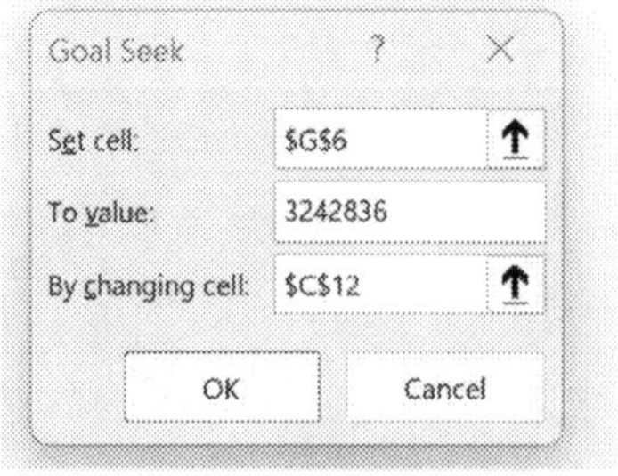

Figure 4.14 Cash Flow 13.

The result would be 7.58%.

You can do this iteration with all other variables. For example, you believe that the discount rate should be 8.0% but then you want to find the exit yield. You can run the same process with Goal Seek but instead of changing the discount rate, you can change the exit yield. The same can be said and done with the growth rate.

Quarterly Discounted Cash Flow

Now that you know the mechanisms for creating and calculating a cash flow, you can use the same technique to build any other cash flow.

The quarterly discounted cash flow only means that instead of having rents, rates, and dates on an annual basis, you will have quarterly rents, quarter dates, and you will have to adjust rates and formulas to deal with quarterly values.

In the property world, rents are typically paid quarterly in advance (UK), and sometimes, even semi-annually in advance in the Middle East. As such, it is more efficient to combine the rent payment profile with the cash flow. As we assume we receive rents quarterly and in advance, we will adopt a quarterly cash flow.

When moving from an annual to quarterly cash flow, things to look out for are the annual growth rates, discount rates, and rents that will need to be adjusted to quarterly figures.

EXERCISE 8

Single Rent Review or Renewal to ERV

Using the same case study, but now we will assume a 10-year holding period and will further assume that the tenant will extend the lease and will have another rent review in year 5.

The investor's target rate is 8.00% for this property assuming all costs of money and risks for this investment.

Layout

The difference between the annual cash flow template and the quarterly cash flow template is that we will need to adjust annual values to quarterly values and vice-versa. We will also need more periods because in an annual cash flow, each period refers to one year; whereas in a quarterly cash flow, each period refers to a quarter, so we will need four times as many columns to calculate the cash flow.

Inputs

1 **Holding period**
2 Since there are four quarters in a year, the number of periods for a quarterly cash flow is:
3 = Years × 4
4 I understand that 4 is a bit like hardcoding, but I can live with that since I haven't yet lived a year, in which the number of quarters in a year is different than 4.
5 Adjust the formatting to show "**qtrs**" instead of "yrs".
6 **Passing rent/ERV**
7 This is the quarterly rent:
8 = Annual Rent/4
9 **Review period**
10 This is the input for when the review period will take place. In this case, the review will take place 3 years from now:
11 = Review Period in Years × 4
12 Change the format to "**qtrs**" as well.
13 **Rental growth and discount rate**
14 We will need to make the annual rates into quarterly rates by de-compounding the rates:
15 Quarterly rate = (1 + annual rate) ^ 0.25 – 1
16 The **Inputs** table will now look such as:

Quarterly Discounted Cash Flow

INPUTS	Annual	Quarterly
Valuation Date	31-Dec-23	
Holding Period	10 yrs	40 qtrs
Passing Rent	215,000	53,750
ERV	273,690	68,423
Review Period	5 yrs	20 qtrs
Rental Growth	2.00%	0.50%
Initial Yield	6.63%	
Exit Yield	8.83%	2.14%
Discount Rate	8.00%	1.94%
Purchase Price	3,242,836	
Sale Price	3,778,330	

Figure 4.15 - Cash Flow 14

Figure 4.15 Cash Flow 14.

Quarterly Discounted Cash Flow

INPUTS	Annual	Quarterly
Valuation Date	45291	
Holding Period	10	=C5*4
Passing Rent	215000	=C6/4
ERV	273690	=C7/4
Review Period	5	=C8*4
Rental Growth	0.02	=(1+C9)^0.25-1
Initial Yield	0.0663	
Exit Yield	0.0883	=(1+C11)^0.25-1
Discount Rate	0.08	=(1+C12)^0.25-1
Purchase Price	=C6/C10	
Sale Price	=C7*(1+C9)^C5/C11	

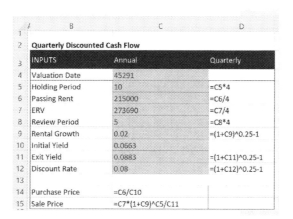

Figure 4.16 Cash Flow 15.

Cash Flow

1. **Period**
2. We can use the **Fill** series again to get all 20 periods in one go.
3. **Date**
4. We will still use the EDATE function but instead of 12 months, we will use 3 months as the second argument for the function.
5. = **EDATE** (start date, 3)
6. The reason for 3 is because there are 3 months in each quarter. Again, it's okay to hardcode it as I haven't yet seen any quarter with more or fewer months than 3.
7. Drag the formula to the right up to end of cash flow.
8. **Purchase price**
9. You just need to link this cell with the Purchase Price input cell.
10. **Rent**
11. The technique that we are going to start using here is one that we will enter the passing rent in Period 0 and then create the formula for rents from period 1.
12. To create the rent formula, we will start using the function = **IF**.
13. The **IF** function will return a value in case the condition is TRUE, and another value in case the condition is FALSE.
14. The Syntax of **IF** function in Excel is:
15. = **IF** (condition, result if true, result if false)
16. In our specific case, the **IF** function will be:
17. = **IF** (cash flow period = review period, ERV × (1 + growth) ^ period, previous rent)
18. In English, this formula would be read somewhat like this:

> If the period in the cash flow is a period in which a review must happen, then I want to compound my market rent by the rental growth over the period, otherwise I want the rent from the previous period.

19. After you calculated the first rent formula in Period 1, you will need to copy the formula to the right. Remember though that you will need to use the dollar sign to fix the cells accordingly.
20. Note that you do not want to have a rent value in the last period of the cash flow, because this cash flow is for a quarterly in *advance* rent receipt; therefore, you will receive the rent in period 0 and not the rent of the last period, which will then be received by the next buyer of the property.
21. **Sale price**
22. In the previous exercise, we just added the sale price directly to the exit period. However, this doesn't follow best practice since we want to create a model that can cope with changes in the input data set. If we changed the holding period from 5 to 4, for example, we would like to see this change appear automatically in cash flow, without the need to change it manually every time we change the holding period.
23. We can do that by adding an IF function from Period 0:
24. = **IF** (cash flow period = holding period, Sale Price, 0)
25. Then, we can copy the formula all the way until the end of the cash flow. You will see the sale price 'appear' at the end of the holding period.
26. **NCF**
27. Nothing new here: the same SUM of everything above (bar periods and dates, of course).

28 The complete cash flow will now be as follows:

	B	C	D	E	F	G	H	V	W	X
6										
7										
8	Cash Flow									
9	Period	0	1	2	3	4	5	19	20	21
10	Date	31-Dec-23	31-Mar-24	30-Jun-24	30-Sep-24	31-Dec-24	31-Mar-25	30-Sep-28	31-Dec-28	31-Mar-29 3
11	Purchase Price	-3,242,836								
12	Rent	53,750	53,750	53,750	53,750	53,750	53,750	53,750	75,544	75,544
13	Sale Price	-	-	-	-	-	-	-	-	-
14	NCF (Net Cash Flow)	-3,189,086	53,750	53,750	53,750	53,750	53,750	53,750	75,544	75,544

Figure 4.17 Cash Flow 16.

29 The formulas will be like:

	B	C	D	E
17				
18	Cash Flow			
19	Period	0	1	2
20	Date	=EDATE($C$4,C19*3)	=EDATE($C$4,D19*3)	=EDATE($C$4,E19*3)
21	Purchase Price	=-C14		
22	Rent	=$D$6	=IF(D19=$D$8,$D$7*(1+$D$9)^D19,C22)	=IF(E19=$D$8,$D$7*(1+$D$9)^E19,D22)
23	Sale Price	=IF(C19=$D$5,$C$15,0)	=IF(D19=$D$5,$C$15,0)	=IF(E19=$D$5,$C$15,0)
24	NCF (Net Cash Flow)	=SUM(C21:C23)	=SUM(D21:D23)	=SUM(E21:E23)

Figure 4.18 Cash Flow 17.

Outputs

1 **IRR**
2 You know how to calculate the IRR:
3 = IRR (Values)
4 The problem with this IRR is that it is a quarterly rate and not an annual rate. Remember that the IRR function will always result in a rate that is of the same periodicity as the cash flow. As we are now in a quarterly cash flow, this IRR will be quarterly.
5 Therefore, we need to compound it up to four quarters in order to get an annual IRR:
6 = (1 + IRR (Values)) ^ **4** – 1
7 Note: we will use the more 'modern' XIRR function further in the book. However, for now, I need you to know how to annualise a quarterly IRR. Remember, it is ^4 and not x4, because the IRR is a growth rate and therefore should be compounded (power of/exponential).
8 **NPV**
9 For the NPV, we will use the same NPV function and syntax as before. We need to remember that the discount rate needs to be consistent with the periodicity of the cash flow and that Value 0 (zero) of our cash flow is not discounted because it's today's value. Therefore, we need to exclude this value from the NPV range, but add it back after the NPV function.
10 = NPV (**qtrly** discount rate, Value 1: Value N) + Value at time 0
11 **Worth/investment value**
12 You will be happy to hear that once you have calculated the NPV, calculating worth is again simply adding the NPV to the purchase price.
13 Worth = NPV + Purchase Price
14 **Profit**

15 The profit calculation is a new item in our Output Table and it is the sum of all the NCF values:
16 Profit = NCF at period 0 + NCF at period 1 + NCF at period 2 + … + NCF at period N
17 In Excel, it would be:
18 = SUM (NCF$_0$: NCF$_n$)
19 **Equity multiple**
20 The only new ratio added is the equity multiple (EM), which is by how much one unit of currency has grown over the holding period. For example, if the EM is 1.8, it means that each dollar, pound, euro, etc. you have invested will come back to you as 1.8. Your profit is therefore 0.8 for each unit, or 80%.
21 EM is a good ratio to be looked at; however, the EM doesn't consider the time value of the money. So, you will need to ask yourself, is 1.8 good over 5 years? Normally, investors will have their own EM benchmarks, for example, 2.0x for 10 years. This means that they expect their money to double every ten years.
22 EM = Profit/Equity + 1
23 As we have already calculated the profit from the previous exercise, all you need to look is the equity side. In this case, it is the purchase price.
24 The '+1' in this case is to account for the equity you have put in the profit calculation.
25 EM = Profit/Purchase Price + 1
26 **Initial yield on worth**
27 The output initial yield on investment value/worth:
28 IY = Annual Passing Rent/Worth
29 **Output table**

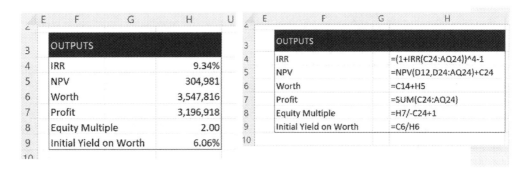

Figure 4.19 Cash Flow 18.

Review Cycles

EXERCISE 9

Now, we are interested in adding a second review to the cash flow. So, we will use a hypothetical case that there will be a review in three years' time and then again in eight years' time, i.e., the review cycle is for every five years. All previous assumptions remain valid for this case.

Layout

We will use exactly the same layout as before, but we will add a row for review cycle. I have highlighted in rows 8,9 for the review cycle and the rent calculation row (23) which we will be focusing on.

Introduction to Discounted Cash Flow Modelling and Analysis   57

	B	C	D	E	F	G	H	U	V	W
1										
2	Quarterly Discounted Cash Flow									
3	INPUTS	Annual	Quarterly		OUTPUTS					
4	Valuation Date	31-Dec-23			IRR		1.71%			
5	Holding Period	10 yrs	40 qtrs		NPV		-1,438,988			
6	Passing Rent	215,000	53,750		Worth		1,803,848			
7	ERV	273,690	68,423		Profit		589,245			
8	First Review Period	3 yrs	12 qtrs		Equity Multiple		1.18			
9	Review Cycle	5 yrs	20 qtrs		Initial Yield on Worth		11.92%			
10	Rental Growth	2.00%	0.50%							
11	Initial Yield	6.63%								
12	Exit Yield	8.83%	2.14%							
13	Discount Rate	8.00%	1.94%							
14										
15	Purchase Price	3,242,836								
16	Sale Price	3,778,330								
17										
18										
19	Cash Flow									
20	Period	0	1	2	3	4	5	18	19	20
21	Date	31-Dec-23	31-Mar-24	30-Jun-24	30-Sep-24	31-Dec-24	31-Mar-25	30-Jun-28	30-Sep-28	31-Dec-28
22	Purchase Price	-3,242,836								
23	Rent		53,750							
24	Sale Price									
25	NCF (Net Cash Flow)	-3,189,086	0	0	0	0	0	0	0	0

Figure 4.20 Cash Flow 19.

Inputs

The only inputs that will change in this case are: the review cycle and adjusting the sale price to be a function of rental income at the time of sale and not based on the ERV at the time of sale.

Review Periods

As we assume a lease agreement with a term of 15 years, whereby rent reviews occur every 5 years, it means that reviews will take place in year 5 and 10 from the date of signing the lease. Because this lease was signed two years ago, the next review will occur in three years' time from now and then in eight years' time from now. Therefore:

First Review Period = 3 years = 12 quarters
Review Cycle = 5 years = 20 quarters

Remember to copy the formula of quarterly input down from Review Period 1 to Review Cycle.

Cash Flow

The rent formula will now have to take into account the second review period. In this case, we can use the Excel function **OR:**
   The **OR** syntax in Excel is:
   = OR (test, test, test, test, ……, test)
   A test is a condition that can either be true or false. So, you will always need a mathematical sign in this test, namely =, >, <, <>. If any of these tests turn out true, the result of the formula will be true.
   In our case, the **OR** function will be formulated as:
   = OR (cash flow period = review period 1, cash flow period = review period 1 + review cycle)
   Adding the new 'review period 2' with the OR function, the rent formula will be:

= IF (**OR (cash flow period = first review period, cash flow period = review period 1 + review cycle),** ERV × (1 + market growth) ^ review cycle, previous rent)

The complete cash flow will now be as follows:

	B	C	D	E	F	G	H	I
19	Cash Flow							
20	Period	0	1	2	3	4	5	6
21	Date	31-Dec-23	31-Mar-24	30-Jun-24	30-Sep-24	31-Dec-24	31-Mar-25	30-Jun-25
22	Purchase Price	-3,242,836						
23	Rent		53,750	53,750	53,750	53,750	53,750	53,750
24	Sale Price	-	-	-	-	-	-	-
25	NCF (Net Cash Flow)	-3,189,086	53,750	53,750	53,750	53,750	53,750	53,750

Figure 4.21 Cash Flow 20.

	B	C	D
19	Cash Flow		
20	Period	0	1
21	Date	31-Dec-23	31-Mar-24
22	Purchase Price	-3,242,836	
23	Rent	53,750	=IF(OR(D20=$D$8,D20=$D$8+$D$9),$D$7*(1+$D$10)^D20,C23)
24	Sale Price	-	-
25	NCF (Net Cash Flow)	-3,189,086	53,750

Figure 4.22 Cash Flow 21.

Outputs

Output Table

	E	F	G
2			
4		OUTPUTS	
5		IRR	13.13%
6		NPV	1,653,088
7		Worth	9,153,088
8		Profit	5,346,408
9		Equity Multiple	1.71
10		Initial Yield	6.67%

Figure 4.23 Cash Flow 22.

	E	F	G
2			
4		OUTPUTS	
5		IRR	=(1+IRR(C24:W24))^4-1
6		NPV	=NPV(D12,D24:W24)+C24
7		Worth	=C14+G6
8		Profit	=SUM(C24:W24)
9		Equity Multiple	=G8/C14 + 1
10		Initial Yield	=C6/C14

Figure 4.24 Cash Flow 23.

COMPLETE MODEL

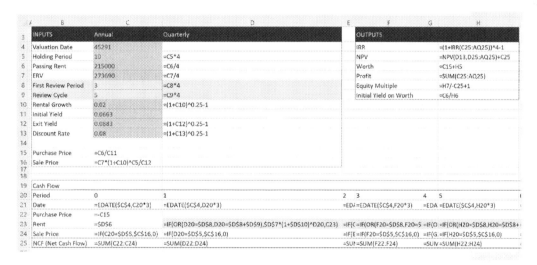

Figure 4.25 Cash Flow 24.

Figure 4.26 Cash Flow 25.

Indexation to Inflation

In the case above, we only had two reviews. What happens when we have annually indexed rent? You could add lots of tests within your **OR** function, but the whole point of financial modelling is to be efficient in how we model formulas. So, let's not do that and think of a better way to achieve the same results.

EXERCISE 10

Let us assume now that the rents are indexed every year. In this case, let us further assume that the rent has just been indexed and that the next indexation will incur in one year's time, in quarter 4.

Layout

I think you may be getting the hang of it by now: the same layout as before.

Inputs

The changes of inputs will mainly incur in Review Period 1, Review Cycle, and ERV. We will also need to discuss the Sale Price forecast a bit.

1 First review and indexation cycle
2 You can change the first review period to 'NA' as it will not be used for now and change review cycle to indexation cycle.
3 You will also need to add a line for inflation rate; I have added it in row 11 after rental growth.
4 The Input table will now look like this:

	A	B	C	D
1				
2		Quarterly Discounted Cash Flow		
3		INPUTS	Annual	Quarterly
4		Valuation Date	31-Dec-23	
5		Holding Period	10 yrs	40 qtrs
6		Passing Rent	215,000	53,750
7		ERV	273,690	68,423
8		First Review Period	NA	
9		Review Cycle	1 yrs	4 qtrs
10		Rental Growth	2.00%	0.50%
11		Inflation Rate	2.50%	
12		Initial Yield	6.63%	
13		Exit Yield	8.83%	2.14%
14		Discount Rate	8.00%	1.94%
15				
16		Purchase Price	3,242,836	
17		Sale Price	3,778,330	

Figure 4.27 Cash Flow 26.

5 Sale price
6 The discussion here is mainly in terms of the numerator, i.e., the rent that should be used when forecasting the sale price. In previous exercises, we have used ERV and grossed it up by the market growth rate up until the time of sale. This assumption is fine under two scenarios: (1) we expect that there will be a rent review to market rents at the time of sale or (2) we expect it to be a vacant possession (however, the yield would be greater than in scenario 1).
7 In cases where there is a lease agreement in place at the expected exit period, best practice and economic logic would dictate that you use the rent forecast at the exit period.
8 In our case, we will further continue to assume that the exit value or estimated sale price will be a function of the future value of the ERV at the end of the holding period.

Cash Flow

Rent Function Including Initial Review Period and Subsequent Review Cycles

In the cash flow, we will only need to change the rent formula again. Otherwise, you can leave all other formulas as they were calculated before.

The best way to approach multiple reviews in single-tenant cash flow models is to use the **MOD** function. The **MOD** is the remainder of a division. The syntax in Excel is:
= MOD (number, divisor)
In our case, the **number** is the cash flow period we are in, and the **divisor** is the review cycle.
=MOD (cash flow period, review cycle)

Introduction to Discounted Cash Flow Modelling and Analysis 61

From the previous period, we had the rent formula:
= IF (**MOD (cash flow period, indexation cycle) = 0**, previous rent × (1 + annual inflation rate), previous rent)

Note that the inflation rate is applied as an annual rate and not quarterly. The reason for that is because you will be indexing once a year only and you will be indexing the previous rent which will already be grossed up from the previous years.

Figure 4.28 Cash Flow 27.

Figure 4.29 Cash Flow 28.

Output Table

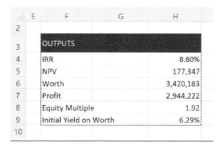

Figure 4.30 Cash Flow 29.

Figure 4.31 Cash Flow 30.

Is It a Good Investment?

So, now that we calculated the cash flow and found the output, we can start thinking of an investment recommendation.

We can base our investment decisions in the above three measures only, i.e., IRR, NPV, and worth.

Your target rate is the main assumption here and should incorporate inflation, inflation risk, country risk, competitive investment opportunities, liquidity risk, propensity for risk, and a lot more. The discussion on the target rate is part of our programme in property risk analysis, available separately. Alternatively, just assume for the time being the target rate is 'right'.

The relationship between target rate, IRR, NPV, and worth is as follows:

Scenario 1: IRR > Target Rate, NPV > 0, Worth > Purchase Price
Scenario 2: IRR = Target Rate, NPV = 0, Worth = Purchase Price
Scenario 3: IRR < Target Rate, NPV < 0, Worth < Purchase Price

So, when do I accept an investment (i.e., I decide to buy)?

In all of them!

Scenario 1: all your risk is being rewarded and a positive NPV means you are making a profit and above your required return. Your maximum bidding price as reflected by the worth is higher than the purchase price.

Scenario 2: all your risk is being reward and there is no super-profit. Worth is, therefore, the same as the purchase price.

Scenario 3: your risk is not being rewarded and you will be making a loss if you decide to buy this at the asking price. Your maximum bidding price is then lower than the purchase price, and if the seller accepts the lower price, it is still a good investment.

Conclusion

In this chapter, we learnt how to create a cash flow from scratch for a single-tenant property. We started with annual cash flow for five years with no reviews and ended with a quarterly cash flow that incorporated lease events such as reviews to market rents and indexation to inflation on an annual basis.

Discounted cash flow models are a great way to approach the fair value discussion as you can explicitly work your way through rent forecast, net cash flow, and risk assessments, for example, the impact of exit yields on returns. The DCF is a rigorous valuation approach and can make you focus on issues that would not be apparent if you are running a conventional valuation only.

Chapter 5

Advanced Discounted Cash Flow Modelling and Analysis

Chapter Contents

Introduction	64
Multi-Let Cash Flows	65
How to Define the Initial and Exit Yields	65
Initial Value (Purchase Price)	68
Exit Value (Sale Price)	68
Holding Period	68
Target Rate/Discount Rate	69
Tenancy Schedule	70
Dating Problem	72
Apportioning Calculation	73
Dating the Cash Flow	73
Growth Series	75
Fixing Cell References	76
Passing Rent	77
Lease Start	77
Lease End	77
Passing Rent	77
First Review to Market Rent	79
Review Date 1	79
Lease End	79
ERV 1	79
Froth 1	79
Second Review to Market Rent	80
Review Date 2	80
Third Review to Market Rent	81
Void, Rent-Free Period, and Second Lease	81
2nd Lease Start	82
End of Rent Free	82
ERV	82
Total Rent Forecast	83
Arrays	83
CAPEX	84
Refurbishment Costs	84

OPEX	85
Void Costs	85
Letting Fees	87
Review Fees	89
Pro-Forma Cash Flow	89
Net Operating Income (NOI)	90
Revenue	90
Net Rents	90
Operating Expenses	90
Net Operating Income	90
Investment Cash Flow	90
Purchase Price	91
Purchaser's Costs	91
Refurbishment CAPEX	91
Sale Price/Exit Value	91
Net Cash Flow	91
Holding Period	91
Revenue Holding Period Switch	91
Operating Expenses and Investment Cash Flow	92
Returns Calculation	92
IRR vs XIRR	93
NPV vs XNPV	93
Price vs Value vs Worth	93
Profit	93
Equity/Peak Equity	94
Return on Equity (ROE) and Equity Multiple	94
What Is the Market Value?	97
Creating a Summarised Annual Cash Flow	99
Conclusion	101

Introduction

So far, we have modelled one single tenant and two types of lease events: rent reviews (or renewals), and indexation to inflation.

We are now interested in modelling more complex, realistic situations, for instance, the cash flow of a building with various tenants, each paying their rents on different dates each quarter, each having lease expiries in different years, and a lot more.

In the examples that will follow, we will use the standard lease terms with adjustments to market rents. This means that for each unit in a building, we will have a rent agreement in place, and this rent will be fixed until the next review period.

We will assume an office building in the UK with a typical lease term for 20 years with a 5-year review cycle to the market rent (or estimated rental value – ERV). You will see that this model structure can be used for shorter lease terms, for example, in the Middle East, where tenants and owners tend to agree a lease of 3 to 5 years with fixed rents, but then renewed at a new market rent at the end of the term.

Multi-Let Cash Flows

Model Structure

As you know by now, the building of models that are elegant and simple is a very complex task. To obtain elegance and simplicity, we need to structure the model in such a way that is visually easy to understand for the outside viewer.

For the purpose of this book, I will model it all in one single sheet, so it is easier to follow. However, you may want to split the calculations in different sheets as printing is much easier if each sheet contains only one printable output.

The below is an example of a model structure which we will adopt in this chapter.

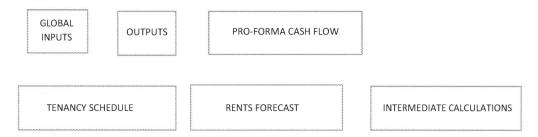

Figure 5.1 Multi-Let Cash Flow Structure.

Global inputs

Global inputs are those that will affect the whole cash flow. For instance, growth rates, exit yield, and any other parameters which will be applied to the cash flow in its entirety, i.e., not just one tenant.

How to Define the Initial and Exit Yields

In the single-let DCF, we considered the purchase and sale price (exit value) a function of the initial and exit yields, and we assumed that these were given by market comparison. However, we know that the initial and exit yields are a function of the relationship between a risk-free rate, risk premium, and expected rental growth. As these are very important concepts in real estate investment valuation, I will break down the initial and exit yields as:

$$\text{Yield} = \text{Risk-Free Rate} + \text{Risk-Premium} - \text{Rental Growth}$$

Risk-Free Rate

The initial yield is a function of the risk over the holding period, as such the risk-free rate should be based on the yield curve. In the UK, it would be the Gilts (see below).

The theory is that because an investor would be 'indifferent' to holding Government bonds or buying a property that would return excess returns (risk premium) that would compensate them for holding the higher risk asset, then the starting point should always be the risk-free rate. As such, the term of the holding period and the term of the bond should match and, in our case, it will be the ten-year bond.

For the initial yield, relying on the spot curve as seen below is a relatively straightforward exercise. However, choosing a ten-year yield that would be applicable for the exit yield in ten years' time is a more difficult task. We tend to use what has been the long-term ten-year bond yields

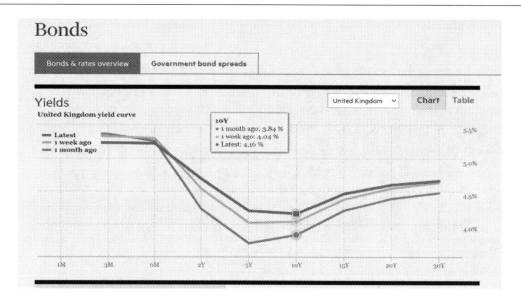

Figure 5.2 The Yield Curve.
Source: ft.com, 14 February 2024.

over the past years with the assumption that the 'past is the best indicator of the future', with the typical financial caveat that 'past performance does not guarantee future results'.

IS THE FORWARD CURVE FOR THE 10TH YEAR A RELIABLE BENCHMARK FOR THE RISK-FREE RATE AT EXIT?

The forward curve is the interpolation of the yield curve, and I am not going to derive the forward curve here, but tell you that what it does is it shows the expected short-term rates – in the UK example, the three-month SONIA. Which means that, if you believe at the time of exit, the yield will be flat, i.e., there will be no variation in rates for different terms, then the forward curve can be used as a forecast risk-free rate for the ten-year bond. However, in the more realistic and economically rational scenario, where the yield curve will be an upward sloping curve, then the forward curve is not the expectation of the 10-year risk-free rate. Moreover, the forward curve is the expectation of interest rates today, and it may not deliver as markets can move very quickly see Figure 5.4.

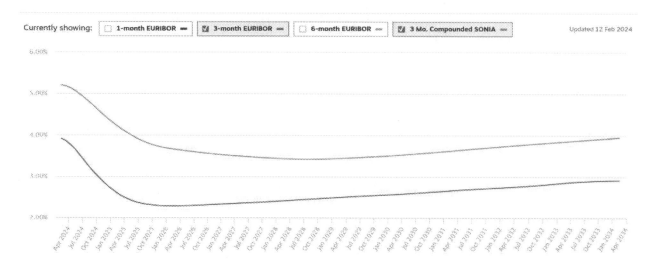

Figure 5.3 Forward Curve 1.
Source: Chatham Financial.

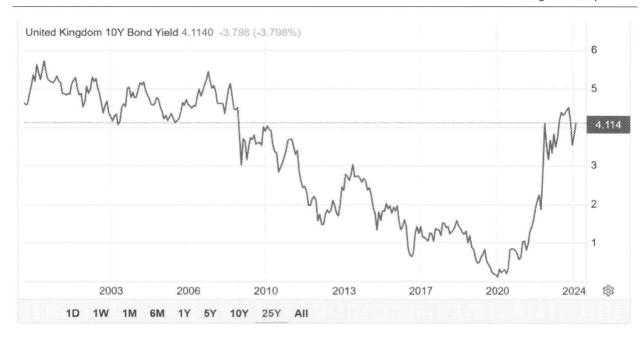

Figure 5.4 Bond Yield 10Y Historic.
Source: Trading Economics.

RISK PREMIUM

Professor Andrew Baum in his latest book, *Real Estate Investment*, Fourth Edition, estimated a value for the expected long-term premium for the property market as 3.0%, measured by delivered returns in the property market in excess of the risk-free rate.

There is a lot that can be said about risk-premiums, but I would suggest that you read Professor Baum's book for the more theoretical approach. In our case, we will assume 3.0% with further adjustments for sector, for example, you could say that a hotel will require a greater risk premium than an office, all things equal (location, quality, size), and we will also need to adjust for location and property specific.

As such, the 3.0% risk premium is an average for the illiquidity of the asset class. And on that note, it can also be said that short-term investors which require higher liquidity will also weigh this illiquidity more heavily; so for them, this risk premium may well be 4.5% or higher. For a long-term investor, however, less concerned about turning the investment in cash, this liquidity premium may be lower, say 2.0%.

RENTAL GROWTH

Note here that the rental growth should be the expected growth in this particular building, i.e., it should take into account *depreciation*. For example, if the market growth for this particular asset in this location is forecasted to be 3.0% p.a. over the next 10 years, you will need to take a view of the market growth – which considers new buildings – and the specific building. Using Professor's Baum estimates of 0.75%, this gives an average for depreciation; however, some buildings may depreciate quicker than others, say data centre vs residential blocks.

A good illustration of rental growth in new and existing properties can be seen by the graph below by Professor Neil Crosby in the 'Discounted Cash Flow Valuations' paper published by the RICS in November 2023.

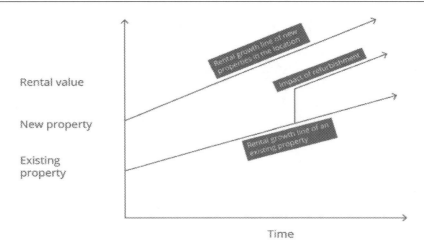

Figure 5.5 Rental Growth with Depreciation.
Source: Discounted Cash Flow Valuations, RICS, November 2023.

Initial Value (Purchase Price)

Gross Value = Total Passing Rent/Net Initial Yield
Net Value = Gross Value/(1 + purchaser's costs)

Exit Value (Sale Price)

It is important to note that the return on investment is very sensitive to the sale price; therefore, this is one area you need to be very careful with. When estimating sale price or exit value, you should anticipate the state of the property, physically and in tenure/leasing terms, at the exit date. This should be overlaid with forecast movements in the general property markets and the expected risk-free rate and risk premium.

Gross Exit Value = Exit Rent/Exit Yield

Therefore, to calculate the sale price, we will look at the exit rent as a function of the ERV at the exit date.

ERV at Exit = ERV x (1 + growth)^holding period
Sale Price = Exit Rent/Exit Yield/(1 + purchaser's cost)

Note that you need to divide the gross exit value by the purchaser's cost because we are assuming that the yields are net. For consistency's sake, if you go in on a net yield basis, you **must** go out on a net yield as well, hence dividing by (1 + purchaser's costs).

If you assume both yields are gross, then your rent/yield will give you the net value, and you will therefore need to add purchaser's costs to the total acquisition value. My point being, do not ignore transaction costs and make sure you understand if the benchmarks for yields are on a gross or net basis.

Holding Period

The holding period will mainly be a function of the business plan – ideally exiting at a period of full stabilisation and at a time when market values are high, and yields are low. However, betting on market movements alone is a high-risk business so the best way to approach the holding period is to exit when the property is stabilised and the weighted average unexpired lease term (WAULT) is long. Typical holding periods would be five or ten years for core and core-plus investments, and shorter for value-add and development investments. If buying or valuing an already stabilised asset, then inevitably the value may deteriorate given the depreciation and obsolescence of the building and the shorter lease terms. This effect can only be reversed if the location benefits from major regeneration and rental growth expectation increases or there is a view that risk-free rates may decrease.

In terms of choosing a holding period for market valuation, the RICS 'Red Book' and the DCF Practice Guidance pretty much leave it to the valuer to choose. I recommend you choose a holding period that is long enough so you can analyse the 'what-ifs' in tenancy cycles; for example, what a void and new tenancy may look like and what kind of capital expenditure may be required for the property to regain a 'stabilised' status. In our case, I left it as ten years as my cash flow runs up to ten years only for the purpose of this book; but we will be able to test different holding periods once the model has been built and the holding period switch is added to the model. If the asset cannot be stabilised up to year 10, then it is important to extend the cash flows further and make annual forecasts until stabilisation can be achieved.

Target Rate/Discount Rate

As mentioned in the *RICS Discounted Cash Flow Valuations Practice Information*, 1st edition, November 2023, there is no consensus concerning the best approach to the determination of the discount rate. Therefore, this is how I approach setting a discount rate for a certain investment.

$$\text{Target or Discount Rate} = \text{Risk-Free} + \text{Risk Premium}$$

Because this is true, some people may be inclined to think that because:

$$\text{Initial Yield} = \text{Risk-Free} + \text{Risk Premium} - \text{Growth}$$

Then:

$$\text{Target or Discount Rate} = \text{Initial Yield} + \text{Growth}$$

Although this may be true, the discount rate can only be determined by Initial Yield + Growth IF and only IF you assume that the exit yield will be THE SAME as the initial yield and the assumed growth is constant over the holding period.

Can We Use WACC as the Discount Rate?

WACC stands for the weighted average cost of capital, and in corporate finance, it is used for firms to decide on their projects, i.e., it is their internal target rate based on their cost of capital: debt and equity. Basically:

WACC = [cost of borrowing × LTV + cost of equity × (1-LTV)]/Total Value

The point I want to make here though is that the *WACC can be used* if you assume that the risk of the project is the same as the average risk in the current portfolio, i.e., the cost of debt and equity will be the same as their current portfolio. For example, if it is a core investor, with minimal gearing, say 20% loan to value (LTV) and the asset being analysed broadly represents the same risk profile, i.e., similar location, condition, sector, tenant covenants and tenancy structure, and the capital structure will follow the same 20% LTV, then using the WACC would be defensible. However, if this core investor now decides to venture onto value-add and opportunistic projects, then the WACC would not make sense.

70 Real Estate Financial Modelling in Excel

	B	C	D	E
8	GLOBAL INPUTS			
10	Date		28-Feb-24	
12	Holding Period		10 yrs	
13	Target Rate		7.50%	
15			Entry	Exit
16	Risk-Free Rate		4.50%	3.00%
17	Risk Premium		3.00%	4.00%
18	Rental Growth		2.00%	1.50%
19	Net Initial Yield		5.50%	5.50%
21	Total Rent		336,000	476,627
22	Gross Value		6,109,091	8,665,942
23	Net Value		5,720,123	8,665,942
25	Letting Fee		15.00%	
26	Review Fee		10.00%	
27	Purchaser's Cost		6.80%	
28	Sale Cost		0.50%	

	A	B	C	D	E
8	GLOBAL INPUTS				
10		Date		45350	
12		Holding Period		10	
13		Target Rate		=D16+D17	
15				Entry	Exit
16		Risk-Free Rate		0.045	0.03
17		Risk Premium		0.03	0.04
18		Rental Growth		0.02	0.015
19		Net Initial Yield		=D16+D17-D18	=E16+E17-E18
21		Total Rent		=SUM(G40:G41)	=I42*(1+D18)^D12
22		Gross Value		=D21/D19	=E21/E19
23		Net Value		=D22/(1+D27)	=E22/(1+E27)
25		Letting Fee		0.15	
26		Review Fee		0.1	
27		Purchaser's Cost		0.068	
28		Sale Cost		0.005	

Figure 5.6 Target Rate.

Tenancy Schedule

These are inputs which are unique to each tenant. For instance, lease start, lease end, new review, rents, and everything else that is tenant specific.

The tenancy schedule below highlights the main inputs that we will need to forecast the rents and costs.

Note that when modelling, I prefer to have this tenancy schedule horizontally, because it is easier to add more tenants this way, which I will show you when we do the case study of a multi-tenanted office building in London. Even though I copied and pasted the tenancy schedule below in two separate figures, they continue horizontally.

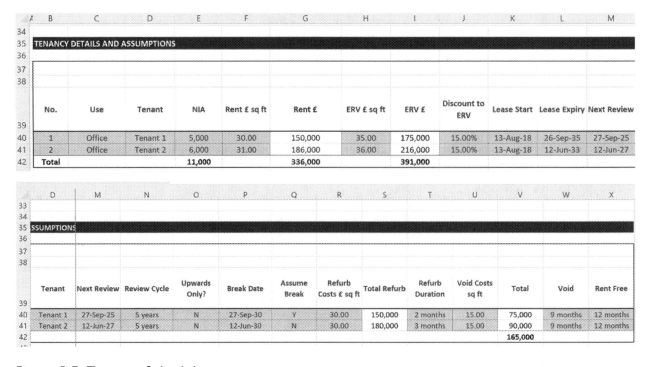

Figure 5.7 Tenancy Schedule.

Advanced Discounted Cash Flow Modelling and Analysis 71

We will go through each of the lease events in detail when we do the modelling, but for the time being, it is important for you to be aware of the structure of the tenancy schedule and how I have structured it in the model.

Output

The output is as before, IRR, NPV, worth, and equity multiple (EM). The outputs are for the entire investment, so you can also say that they are global outputs.

Rents Forecast

AF	AG	AH	AI	AJ	AK	AL	AM
35							
36	Period		0	1	2	3	4
37	Date		28-Feb-24	31-Mar-24	30-Jun-24	30-Sep-24	31-Dec-24
38	End Date		31-Mar-24	30-Jun-24	30-Sep-24	31-Dec-24	31-Mar-25
39	**Tenant**						
40	Tenant 1		14,851	41,768	41,768	41,768	41,768
41	Tenant 2		14,851	41,768	41,768	41,768	41,768
42	**Total**		**29,702**	**83,536**	**83,536**	**83,536**	**83,536**

Figure 5.8 Rents Forecast.

Intermediate Calculations

Remember that financial models should be simple and elegant. We will, therefore, use intermediate calculations to make the formulas, especially the rent formula, as simple as we can. The intermediate calculations will model each lease event separately and we will then add them up so we can arrive at a **Rent Forecast**.

In the intermediate calculations, we will model the:

- Passing Rent; Lease 1 – First Review; Lease 1 – Second Review; Lease 1 – Third Review; Lease 2 – Lease Start; CAPEX; Void 1 – Void Costs; Void 2 – Void Costs; Letting Fees; Review Fees

Pro-Forma Cash Flow

The cash flow will incorporate both cash inflows (rents in our case) as well as cash outlays (capital expenditure, and operating expenses – letting fees, review fees, and void costs).

The cash flow will then give us the net cash flow (NCF) for the property after all expenses and costs were taken into account, but for now, we are not considering any gearing or leverage (debt arrangements) and we are not considering any tax implications.

In this chapter, we will not be adding the debt part in the modelling, which will be closely looked in the next chapters. The important task is to look at the asset alone without any financial engineering involved.

The pro-forma cash flow figure will outline the structure of the cash flow. You will notice that I have left many columns between column AA and column AI, when the cash flow starts. The reason for that is because I wanted to align the dates with the Rents Forecast table below, so column AI is the starting point for both the cash flow and the rents forecast underneath.

Before we can start modelling the rents and the costs, I need to go through some points that are important when modelling cash flows; namely, the dating problem of cash flows, apportionment calculations, and growth series, and I will also make a note on fixing cells.

	AH	AI	AJ	AK	AL	AM
2	**Pro-forma Cash Flow**					
4	Period	0	1	2	3	4
5	Start Date	28-Feb-24	31-Mar-24	30-Jun-24	30-Sep-24	31-Dec-24
6	End Date	31-Mar-24	30-Jun-24	30-Sep-24	31-Dec-24	31-Mar-25
7	**Revenue**	29,867	84,000	84,000	84,000	84,000
8	Net Rent	29,867	84,000	84,000	84,000	84,000
10	**Operating Expenses (OPEX)**	-	-	-	-	-
11	Void Costs	-	-	-	-	-
12	Letting Fees	-	-	-	-	-
13	Review Fees	-	-	-	-	-
15	**Net Operating Income (NOI)**	29,867	84,000	84,000	84,000	84,000
17	**Investment Cash Flow**	(6,109,091)	-	-	-	-
18	Purchase Price	(5,720,123)				
19	Purchaser's Costs	(388,968)				
20	Refurb CAPEX	-				
21	Sale Price	-				
22	Sale Costs	-				
24	**Net Cash Flow (NCF)**	(6,079,224)	84,000	84,000	84,000	84,000

Figure 5.9 Pro-Forma Cash Flow.

Dating Problem

In cash flows, the main dating problems are deciding on:

Frequency of cash flows (annual, semi-annual, quarterly, monthly, daily)
Start and end dates of cash flows
What happens if events happen between start and end dates of cash flows

First, I need to say that for real estate investment, I will model quarterly cash flows and I will also be using the end of the month dates for each quarter, i.e., 31 March, 30 June, 30 September, and 31 December. The reason for needing to say this is because I have seen cash flows being modelled on first days of the month, on 30th of each month, independently of the month being 30, 31, and English quarter dates (25 March, 24 June, 29 September, and 25 December).

The reason I model on a quarterly basis is because rents tend to be paid quarterly in advance in most European countries and property investments tend to be for over three years of hold, so quarterly cash flows may mirror the rental receipts more closely without over-complicating or over-simplifying the model and assumptions.

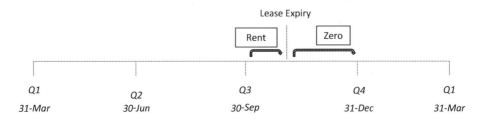

Figure 5.10 Apportionment Example.

Apportioning Calculation

Apportioning means that if an event happens between the start of the quarter and the end of quarter, then the rent or the cost will be calculated on a pro-rata basis. For example, if the lease starts on 15 February and the quarter starts on 1 January and ends on 31 March, the rent will be calculated as,

Rent p.a. × (31 March–15 February)/360

The reason for dividing for 360 is because the rents are normally calculated quarterly; if the rent is 100,000, then each quarter, the rent would be 25,000 without considering if there are more or less days in each quarter.

This is fundamental in this model in that the calculation will always be a function of:

= (End Date – Start Date)/360

Because we want the days based on 360 days, we will use the function in Excel

= DAYS360

So that every quarter the rents are the same, but it also works for month and half-years.

For example, if you think in months, the January will need to have a higher rent than February, because January has more days than February. However, convention dictates that we normally pay or receive the same rent every month. In our case, every quarter.

The syntax for DAYS360 is:

=DAYS360(start date, end date, [method])

The start and end dates seem obvious; for example, if the lease starts on the 15th of February and the cash flow period starts on the 1st of January and ends on 31st of March, the start date for the rent is the 15th of February. The end date is therefore the end of the quarter date.

If it is a lease expiry on the 15th of February, then the start date would be the beginning of the quarter and the end date would be the lease expiry date.

Method is a way of Excel to choose the type of 360 days, because there are European and US 360 days. I chose TRUE, because it gives us consistently equal rents across quarters and months.

As such, our formula for the intermediate calculation will always look like:

= DAYS360 (The Greater of Period Start Date OR Lease Event Date, The Lesser of Period End Date OR Lease Event Date)

Now, the problem is telling Excel how to choose the greater or the lesser of two dates. But as you guessed, Excel can deal with anything, and in this case, it will be the MAX (for maximum or greater of) and MIN (for minimum or lesser of).

Therefore:

=DAYS360(MAX(lease event start, period start), MIN(lease event end, period end), TRUE)

So, in the case of a lease starting on 15 February, in our quarterly cash flow, the calculation would be:

=DAYS360(MAX(15 Feb, 1 Jan), MIN(lease end date, 31 Mar), TRUE)/360 × Rent p.a.

The only problem now is that Excel recognises as negative numbers when leases start AFTER the period end date (for example, 16 April), or when leases end BEFORE the period start date (for example, 1 February). As such, we need to tell Excel, and in this case, the DAYS360 does not work. In order to solve this problem, we need to MAX this formula to ZERO. So, the formula will be:

= MAX(DAYS360 (MAX (Start,End), MIN(Start,End),TRUE), 0)/360 × Rent p.a.

For this model, this will be the foundation for the calculation of each tenant's rents and costs.

Dating the Cash Flow

As mentioned before, my model is quarterly, and we will start the cash flow with the actual valuation date – or acquisition date and the end date is the end-of-quarter date. The aim here is to achieve the end of quarter date from the actual date; for example, if the start date is 15 January 2024, then the end of the quarter will be 31 March 2024. To achieve this, we will need to find which quarter the month 'belongs'. January, February, and March are first quarter; April, May, and June are the second quarter and so on.

Step 1: MONTH/3. So, if Jan (1)/3 = 0.333

Step 2: Approximate to the nearest integer. So, 0.33 would be approximated to 1, which gives us the right answer because January is in Quarter 1.

To achieve this in Excel, we will use the ROUNDUP function:
=ROUNDUP (Number, Decimal)

The number to approximate is the MONTH/3. And decimal will be 0 (zero), because the integer has no decimal places.

The formula will then become:
=ROUNDUP(MONTH(15/01/2024)/3,0)
= ROUNDUP(1/3,0)
= 1

Now, all what we need to do is to multiply the quarter by 3 to arrive to the end-of-quarter month.

Month	Quarter	End of Quarter Month
Jan = 1	1/3 = 0.33 Roundup to 1 (first quarter)	1 x 3 = 3 (March)
Feb = 2	2/3 = 0.66 Roundup to 1 (still first quarter)	1 x 3 = 3 (March)
Aug = 8	8/3 = 2.66 Roundup to 3 (third quarter)	3 x 3 = 9 (September)

Figure 5.11 Finding the Quarter Dates.

Now, we need to put the quarter months in a DATE function:
Understanding the DATE function in Excel:
The DATE function will return the DAY, MONTH, YEAR for any number.
The syntax is as follows:
=DATE(YEAR,MONTH,DAY)
So, in our case, the DATE function will be:
= DATE (YEAR (start), ROUNDUP (MONTH (start)/3, 0) × 3, **30 or 31**)

Now, the only issue is defining the end of quarter to be either 30 or 31. March and December should be 31, whereas June and September should be 30.

To achieve this, we could have added an EOMONTH in the beginning of the formula (before DATE). However, I found out that if you put 0 (zero) for the DAY, it will return the last day of the previous month. As such, I added an '+1' in the MONTH argument so that it goes back to the quarter month.

= DATE (YEAR (start), ROUNDUP (MONTH (start)/3, 0) × 3 **+1, 0**)

The only problem now is that this formula does not work when the start date is an already end-of-quarter date (31/3, 30/6, 30/9, and 31/12). My solution was to resort to the IF function saying:

= IF (Start Date = DATE (that we just calculated with the ROUNDUP), then add 3 months from the Start Date only, otherwise calculate the DATE function).

Advanced Discounted Cash Flow Modelling and Analysis 75

We will use the EOMONTH to add 3 months to the start date.
So, it will be:
= IF(Start Date = DATE(YEAR(Start Date),ROUNDUP(MONTH(Start Date)/3,0) x 3+1,0), EOMONTH(Start Date,3), DATE(YEAR(Start Date),ROUNDUP(MONTH(Start Date)/3,0)*3+1,0))

I know you must be asking why we just don't 'force' the user to enter an end-of-quarter date instead of going through the hoops of creating such a complex formula to start with – but the idea is that because we are apportioning the rents and costs, then using more accurate dates would make more sense.

Then, the next Start Date will be the end of the previous quarter date and the end of the quarter date will be an EOMONTH function:
= EOMONTH(Start Period Date, 3)
And then from then on, you can just copy to the right:

	CK	CL	CM	CN	CO
35					
36	Period	0		=CM36+1	=CN36+1
37	Start Date	=D10		=CM38	=CN38
38	End Date		=IF(CM37=DATE(YEAR(D10),ROUNDUP(MONTH(D10)/3,0)*3+1,0), EOMONTH(CM37,3),DATE(YEAR(D10),ROUNDUP(MONTH(D10)/3,0)*3+1,0))	=EOMONTH(CM38,3)	=EOMONTH(CN38,3)

Figure 5.12 Dating Cash Flows.

Growth Series

Instead of using a fixed growth rate like we did in the single-tenant model, I will now add a growth series. For simplicity, I have left the rates the same for both the market growth as well as for depreciation, but this table can be used if you have different rates for each year.

	L	M	N	O	P
8					
9		GROWTH AND DEPRECIATION ASSUMPTIONS			
10					
11		Year	Rental Growth		
12		Year	Market Growth	Depreciation	Rental Growth
13		2024	3.00%	1.00%	2.00%
14		2025	3.00%	1.00%	2.00%
15		2026	3.00%	1.00%	2.00%
16		2027	3.00%	1.00%	2.00%
17		2028	3.00%	1.00%	2.00%

Figure 5.13 Growth Series.

The rental growth series will need to be calculated on a cumulative basis and increased quarter by quarter. The idea is that the ERV will grow at the rental growth rate.

Insert this rental growth series as an intermediate calculation in your model. This rental growth series will be a geometric growth rate, such as:
$$= (1 + g_1) \times (1 + g_2) \times \ldots \times (1 + g_n)$$
We will use the **LOOKUP** whose syntax is:
= LOOKUP (lookup value, lookup vector, result vector)

For example, if you have the table below and want to find the value of C, you could use the LOOKUP function:

A	10
B	20
C	30

Figure 5.14 LOOKUP.

= LOOKUP (C, Column 1, Column 2) = 30.
In our case, the Multiple Growth Series function will be:
We will start our series in 1 and then:
= Previous Index Value x (1 + LOOKUP (YEAR (Cash Flow Date), Years Range [in forecast table], Rental Growth [in forecast table]) ^ (1/4)

Fixing Cell References

So far, we have only used the absolute reference to fix cells in Excel, but the good news is that Excel also works in a relative reference system. This means that as you copy one formula from one cell to another, the reference will change if you don't fix the cells, i.e. put the dollar signs. In the previous chapter, we used the dollar sign quite frequently, but the technique used was to fix the whole cell, column, and row, adding the dollar sign before the letter and before the number.

Now, we don't just want to fix the whole cell, but the column or the row. If we want to fix the column, the dollar sign goes in front of the letter. If we want to fix the row, the dollar sign goes in the front of the number.

Summarising:

= **$A1**: This means Column A is the fixed-point as the dollar sign is before the Column letter A. If you copy or drag this formula anywhere else on the worksheet, it will always reference back to Column A. The row will change depending on where the formula is copied.

= $A1	= $A1	= $A1
= $A2	= $A2	= $A2
= $A3	= $A3	= $A3

Figure 5.15 Cell Referencing 1.

= **A$1**: This means Row 1 is the fixed-point as the Dollar sign is before the row number 1. If you copy or drag this formula anywhere else on the worksheet, it will always reference back to Row 1. The column will change depending on where the formula is copied to.

= A$1	= B$1	= C$1
= A$1	= B$1	= C$1
= A$1	= B$1	= C$1

Figure 5.16 Cell Referencing 2.

Advanced Discounted Cash Flow Modelling and Analysis 77

= $A$1 .... This means Cell A1 is the fixed-point as there is a Dollar sign before the column letter and the row number. No matter where you copy or drag this formula on the Excel worksheet, it will always refer to Cell **A1**.

= $A1	= $A1	= $A1
= $A1	= $A1	= $A1
= $A1	= $A1	= $A1

Figure 5.17 Cell Referencing 3.

Tip:
My tip is to think like this:

If the cell you are referencing to is on **top** of the cell you are writing the formula in, you want to fix the row. For example, if the formula is in C5 and the reference cell is C1, then you fix the row: add the dollar sign in front of the number, = **C$1**. It is a cash flow-related reference.

If the cell you are referencing to is on the **left** of the cell you are writing the formula in, you want to fix the row. For example, if the formula is in C5 and the reference cell is A5, then you fix the row: add the dollar sign in front of the letter, = **$A5**. It is a tenant-related reference.

If the cell you are referencing to is on **top left** of the cell you are writing the formula in, you want to fix the row AND the column. For example, if the formula is in C5 and the reference cell is A1, then you fix the row: add the dollar sign in front of the number, = **$A$1.** It is a global input.

Using this logic, we can adjust the rent formula before copying it to the right, all the way to period 40.

Passing Rent

We first need to get the information that is relevant to the passing rent, which is start date of the lease, the end date, and the passing rent. I will link the dates and rents to the tenancy schedule.

Lease Start

If there is a tenant in place, then this date will be in the past as per tenancy schedule. If the unit is vacant, then the analyst/valuer will need to estimate a date in the future when they think the new lease will start.

Lease End

The lease end will depend on whether the break is activated or not and it will be an IF function:
= IF (Break= "N", Lease Expiry, Break Date)

Passing Rent

If there is a tenant in place, then the passing rent is known and can be found in the tenancy schedule. However, in case of a vacant unit, the passing rent will be a function of the ERV. However, the ERV is as of the valuation or acquisition date and will therefore need to be grossed up to rental growth.
= IF (Passing Rent is Greater than Zero, Passing Rent, ERV × (LOOKUP (Start Date, Cash Flow Dates, Rental Growth Series)

C	CD	CE	CF	CG	CHCICJ	CK	CL	CM
34						Rental Growth		1.0000
35	Lease 1 - Passing Rent							
36						Period		0
37						Start Date		28-Feb-24
38						End Date		31-Mar-24
39		Tenant	Lease Start	Lease End	Passing	Tenant		
40		Tenant 1	13-Aug-18	27-Sep-30	150,000	Tenant 1		13,333
41		Tenant 2	13-Aug-18	12-Jun-33	186,000	Tenant 2		16,533
42		Total			336,000	Total		29,867
43								

	CC	CD	CE	CF	CG
35					Lease 1 - Passing Rent
36					
37					
38					
39		Tenant	Lease Start	Lease End	Passing
40		=D40	=K40	=IF(Q40="N",L40,P40)	=IF(G40<>0,G40,I40*LOOKUP(CE40,$CM$37:$EE$37,$CP$34:$EE$34))
41		=D41	=K41	=IF(Q41="N",L41,P41)	=IF(G41<>0,G41,I41*LOOKUP(CE41,$CM$37:$EE$37,$CP$34:$EE$34))
42		Total			=SUM(CG40:CG41)

Figure 5.18 Passing Rent.

Now that we found the main inputs for the passing rent, we need to model the actual rent. We will use the DAYS360/MAX/MIN function as discussed before.

The passing rent function will be:

= (MAX(DAYS360(MAX(Lease Start Date, Cash Flow Start Date),

MIN(Lease End Date, Cash Flow End Date), TRUE),0))/360 × Passing Rent

As you see the rent for Tenant 1 is apportioned in the last quarter of the lease end date because the lease end is 30 September, but the lease end is 27 September, so the formula extracts the extra three days.

	CHCICJ	CK	CL	CM	CN	CO	CP	CQ	CR	CS	CT	DM	DN	DO
33														
34		Rental Growth		1.0000	1.0050	1.0100	1.0150	1.0200	1.0251	1.0301	1.0353	1.1374	1.1430	1.1487
35														
36		Period		0	1	2	3	4	5	6	7	26	27	28
37		Start Date		28-Feb-24	31-Mar-24	30-Jun-24	30-Sep-24	31-Dec-24	31-Mar-25	30-Jun-25	30-Sep-25	30-Jun-30	30-Sep-30	31-Dec-30
38		End Date		31-Mar-24	30-Jun-24	30-Sep-24	31-Dec-24	31-Mar-25	30-Jun-25	30-Sep-25	31-Dec-25	30-Sep-30	31-Dec-30	31-Mar-31
39		Tenant												
40		Tenant 1		13,333	37,500	37,500	37,500	37,500	37,500	37,500	37,500	36,250	0	0
41		Tenant 2		16,533	46,500	46,500	46,500	46,500	46,500	46,500	46,500	46,500	46,500	46,500
42		Total		29,867	84,000	84,000	84,000	84,000	84,000	84,000	84,000	82,750	46,500	46,500

	CK	CL	CM
31			
32			
33			
34	Rental Growth	1	
35			
36	Period	0	
37	Start Date	=D10	
38	End Date	=IF(CM37=DATE(YEAR(D10),ROUNDUP(MONTH(D10)/3,0)*3+1,0), EOMONTH(CM37,3),DATE(YEAR(D10),ROUNDUP(MONTH(D10)/3,0)*3+1,0))	
39	=CD39:CD42		
40		=(MAX(DAYS360(MAX($CE40,CM$37),MIN($CF40,CM$38), TRUE),0))/360*$CG40	
41		=(MAX(DAYS360(MAX($CE41,CM$37),MIN($CF41,CM$38), TRUE),0))/360*$CG41	
42		=SUM(CM40:CM41)	

Figure 5.19 Passing Rent 2.

First Review to Market Rent

The review to market rent will happen at the review date – or renewal date if leases are short and fixed, but there is an expectation that it is renewed at the end of the lease term. The inputs that we need to adjust the rent to market rent are Review Date 1, Lease End, ERV 1, and Froth 1.

Review Date 1

This may already be in the tenancy schedule. However, if the lease will end before the review date because the break happens before the lease is reviewed or simply because the lease will expire and there are no interim reviews, then this Review Date 1 will be an empty cell. Note that empty cells in Excel are treated as double quotation marks or " ".

Also, if the unit is initially vacant, then the next review date will be the estimated lease start date plus the number of years for review cycles. For example, if the start date is in one year from now and the review cycle is every five years, then the next review date is in six years' time.

In Excel, these conditions are expressed as an IF function:

= IF(OR(Next Review Date>= Lease End, Next Review = empty ""), nothing "", MAX(Next Review, EDATE(Lease Start, Review Cycle × 12)))

Lease End

We will just link it to the lease end that we calculated for the passing rent.

ERV 1

The ERV is the market rent grossed up by the rental growth, i.e., the future value of the ERV as per tenancy schedule input. However, you will need to note that when there is a review to market, the tenant will need to look at comparables and will need to appreciate the difference between headline rent and net effective rent. In our case of rent review, we should consider the net effective ERV, i.e., discounting for tenant incentives such as rent-free periods and capital contributions that may be included in headline rents.

You could be more precise in this calculation if you know the incentives embedded in a headline rent. For example, if there is a 12-month rent-free period in a 5 (five) year contract and if you use an 8.0% discount rate, the NPV of the 4 years rent is 3.07 and the annuity (PMT) of 3.07 for 5 years is 0.77. Therefore, the discount to headline ERV would have been 23%.

In our case, I just added a % as discount to ERV to calculate the net effective ERV and our formula is:

=IF (Review Date = "" (empty), 0 (zero), ERV × (1 – discount to ERV) × LOOKUP (Review Date, Cash Flow Dates, Rental Growth Series)

Froth 1

The froth is the ERV – passing rent. This froth can however be positive or negative. If the net effective ERV is greater than the passing rent, then this will be positive and vice-versa. However, there is also an 'Upwards-Only' clause in the UK, whereby rents cannot go down. If that is the case, then the froth will never be negative. So, our formula will be:

=IF(Review Date 1= "" (empty),0 (zero), IF(Upwards Only="Y" (Yes), MAX(ERV – Passing Rent,0), ERV – Passing Rent))

80 Real Estate Financial Modelling in Excel

EG	EH	EI	EJ	EK	EL
34					
35	Lease 1 - First Review				
36					
37					
38					
39	Tenant	Review Date 1	Lease End	ERV 1	Froth 1
40	Tenant 1	27-Sep-25	27-Sep-30	160,892	10,892
41	Tenant 2	12-Jun-27	12-Jun-33	214,198	28,198
42	Total			375,090	39,090

	EH	EI	EJ
34			
35	Lease 1 - First Review		
36			
37			
38			
39	=$CD39	Review Date 1	Lease End
40	=$CD40	=IF(OR(M40>=EJ40,M40=""),"",MAX(M40,EDATE(CE40,$N40*12)))	=CF40
41	=$CD41	=IF(OR(M41>=EJ41,M41=""),"",MAX(M41,EDATE(CE41,$N41*12)))	=CF41
42	Total		

	EK	EL
35		
36		
37		
38		
39	ERV 1	Froth 1
40	=IF(EI40="",0,$I40*(1-J40)*LOOKUP(EI40,$EQ$37:$GI$37,$EQ$34#))	=IF(EI40="",0,IF(O40="Y",MAX(EK40-$CG40,0),EK40-$CG40))
41	=IF(EI41="",0,$I41*(1-J41)*LOOKUP(EI41,$EQ$37:$GI$37,$EQ$34#))	=IF(EI41="",0,IF(O41="Y",MAX(EK41-$CG41,0),EK41-$CG41))
42	=SUM(EK40:EK41)	=SUM(EL40:EL41)

Figure 5.20 Review 1.

Now, the formula is the same as for the passing rent; the difference now is that the start, end, and rent will be the Review Date 1, Lease End, and Froth, respectively:

= (MAX(DAYS360(MAX(Review Date 1, Cash Flow Start),MIN(Lease End, Cash Flow End) TRUE),0))/360 × Froth 1

= (MAX(DAYS360(MAX($EI40, EQ$37), MIN($EJ40, EQ$38), TRUE),0))/360*$EL40

Second Review to Market Rent

The process here is the same as for the Review Date 1; the only difference between them is that the Review Date 2 will be calculated from the first review date plus the number of years in the review or renewal cycles. The inputs that we will need are the Review Date 2, Lease End, ERV 2, and Froth 2.

Review Date 2

The Review Date 2 will be the First Review Date + Years of Review Cycles. To achieve this, we will use the EDATE function.

=EDATE (Review Date 1, Review Cycle × 12)

However, if the Review Date 2 is after the Lease End Date, then there will be no Review Date 2. As such, we will need to add the IF function:

= IF(Review Date 1="","",
IF(EDATE(Review Date 1, Review Cycle × 12)>= Lease End Date, "", EDATE(Review Date 1, Review Cycle × 12)))

Lease End, ERV 2 and Froth 2 follow the same logic as in the First Review to Market.

	GK	GL	GM	GN	GO	GP
35		Lease 1 - Second Review				
36						
37						
38						
39		Tenant	Review Date 2	Lease End	ERV 2	Froth 2
40		Tenant 1		27-Sep-30	0	0
41		Tenant 2	12-Jun-32	12-Jun-33	225,685	11,487
42		Total			225,685	11,487

Figure 5.21 Review 2.

	GL	GM	GN	GO	GP
35			Lease 1 - Second Review		
36					
37					
38					
39	=$CD39	Review Date 2	Lease End	ERV 2	Froth 2
40	=$CD40	=IF(EI40="","", IF(EDATE(EI40,$N40*12)>=$GN40,"", EDATE(EI40,$N40*12)))	=EJ40	=IF(GM40="",0, $I40*(1-J40)*LOOKUP(GM40,$GT$37:$IM$37, $GU$34#))	=IF(GM40="",0,IF(O40="Y", MAX(GO40- MAX(CG40,EK40),0),GO40-EK40))
41	=$CD41	=IF(EI41="","", IF(EDATE(EI41,$N41*12)>=$GN41,"", EDATE(EI41,$N41*12)))	=EJ41	=IF(GM41="",0, $I41*(1-J41)*LOOKUP(GM41,$GT$37:$IM$37, $GU$34#))	=IF(GM41="",0,IF(O41="Y", MAX(GO41- MAX(CG41,EK41),0),GO41-EK41))
42	=$CD42			=SUM(GO40:GO41)	=SUM(GP40:GP41)

Figure 5.22 Review 3.

The formula for the Froth 2 follows the same logic:
= (MAX(DAYS360(MAX(Review Date 2, Cash Flow Start), MIN(Lease End, Cash Flow End) TRUE),0))/360 × Froth 2
= (MAX(DAYS360(MAX($GM40, GU$37), MIN($GN40, GU$38), TRUE),0))/360*$GP40

Third Review to Market Rent

The concept of reviews to market rents are the same, and if you need to add a third, fourth, and even a fifth review, the logic will be absolutely the same as for the second review to market rents.

Void, Rent-Free Period, and Second Lease

Once the first lease has ended – either through a break or lease expiry – we will assume that a second lease will take place. In our model, I will assume only two sets of leases, but if you are modelling shorter lease terms, say three years, and your holding period is for ten years, then you should model three sets of leases, because we are interested in knowing the economic value of the property and what a new tenancy would look like after a period of vacancy. As such, I will assume that the 2nd lease will continue until the end of the holding period.

The inputs that we will need for the second lease are the assumed 2nd Lease Start Date, End of the Rent-Free Period Date, and the new ERV.

2nd Lease Start

The 2nd Lease Start Date will be the 1st Lease End Date + Period of Void
As such, we will use the EDATE function:
= EDATE (Lease End Date, Void Period)

End of Rent Free

We will now assume that there will be a tenant incentive in the form of rent-free period, whereby the tenant will not pay rent but will pay the maintenance costs, including insurance and occupancy taxes, such as business rates as they are referred to in the UK.

The End of Rent-Free Date will be the 2nd Lease Start + Period of Rent Free
Using the EDATE function as well:
= EDATE (2nd Lease Start, Rent-Free Period)

ERV

In our case, we will use the ERV as estimated initially. However, there may be cases where a new ERV can be estimated when the business plan will assume an extensive refurbishment and the property may be leased at a higher rent than in the original condition of the building.

The ERV will be the day 1 ERV but grossed up by the rental growth up to the start of the 2nd lease. If you are using a new ERV, say post-refurbishment ERV, you may also want to add a new growth assumption that would consider a lesser depreciation rate because the building is in a better condition.

The new ERV will therefore be:
= ERV × LOOKUP (2nd Lease Start, Cash Flow Dates, Rental Growth Series)
For ease of following, I have linked the growth series from the previous ERV calculation.

	KT	KU	KV	KW
34				
35	Lease 2 - Lease Start			
36				
37				
38				
39	Tenant	2nd Lease Start	End of Rent Free	ERV
40	Tenant 1	27-Jun-31	27-Jun-32	212,455
41	Tenant 2	12-Mar-34	12-Mar-35	269,504
42	Total			481,959

	KT	KU	KV	KW
34				
35	Lease 2 - Lease			
36				
37				
38				
39	=$CD39	2nd Lease Start	End of Rent Free	ERV
40	=$CD40	=EDATE(IR40,W40)	=EDATE(KU40,X40)	=I40*(LOOKUP(KU40,$LC$37:$MU$37,$LC$34:$MU$34))
41	=$CD41	=EDATE(IR41,W41)	=EDATE(KU41,X41)	=I41*(LOOKUP(KU41,$LC$37:$MU$37,$LC$34:$MU$34))
42	=$CD42			=SUM(KW40:KW41)

Figure 5.23 Review 4.

The formula for the 2nd Lease will be:

= (MAX(DAYS360(MAX(End of Rent Free, Cash Flow Start),MIN(Exit Date, Cash Flow End) TRUE),0))/360 × ERV2

= (MAX(DAYS360(MAX($KV40, LD$37),MIN(EDATE(exit_date,12),LD38), TRUE),0))/360*$KW40

Note that even though the lease starts earlier, the rent will only be received at the end of the rent-free period; however, the ERV will be the one agreed at the start of the lease and not the estimated ERV at the end of the rent-free period.

Total Rent Forecast

Now that we have all the components to forecast the rents over the holding period, all what we need to do is add them all up. Easy! You will see that I add them up in a table in the beginning of the sheet, right next to the tenancy schedule so we can then align the cash flow calculation above and it will be easier to see the impact of any changes in the tenancy schedule in the rent forecast.

		AG	AH	AI	AJ	AK	AL	AM	BN	BO	BP	BQ	BR
34													
35	Rents Forecast												
36			Period	0	1	2	3	4	31	32	33	34	35
37			Start Date	28-Feb-24	31-Mar-24	30-Jun-24	30-Sep-24	31-Dec-24	30-Sep-31	31-Dec-31	31-Mar-32	30-Jun-32	30-Sep-32
38			End Date	31-Mar-24	30-Jun-24	30-Sep-24	31-Dec-24	31-Mar-25	31-Dec-31	31-Mar-32	30-Jun-32	30-Sep-32	31-Dec-32
39	Tenant		Tenant										
40	Tenant 1		Tenant 1	13,333	37,500	37,500	37,500	37,500	0	0	1,770	53,114	53,114
41	Tenant 2		Tenant 2	16,533	46,500	46,500	46,500	46,500	53,549	53,549	54,124	56,421	56,421
42			Total	29,867	84,000	84,000	84,000	84,000	53,549	53,549	55,894	109,535	109,535

	AF	AG	AH	AI	AJ	AK
34						
35						
36		Period		=VW36#		
37		Start Date				
38		End Date				
39	=Z39					
40	=Z40			=CM40+EQ40+GU40+IY40+LC40	=CN40+ER40+GV40+IZ40+LD40	=CO40+ES40+GW40+JA40+LE40
41	=Z41			=CM41+EQ41+GU41+IY41+LC41	=CN41+ER41+GV41+IZ41+LD41	=CO41+ES41+GW41+JA41+LE41
42	Total			=SUM(AI39:AI41)	=SUM(AJ39:AJ41)	=SUM(AK39:AK41)

Figure 5.24 Review 5.

Arrays

You will notice in the Period formula, that it refers to VW36# and the adjacent cells are all empty, but when you see the values of the total rent forecast, you will realise that they are all (correctly) populated. This is the 'new' Array function in Excel, in which, instead of selecting one cell and dragging or copying to the right and/or down, you can select the whole range you want to link and Excel will do it automatically for you. This is wonderful! So, why didn't I use it more extensively before? The reason for that is this 'new' function only works in Microsoft 365 and in PC versions. If you are using a Macintosh or using Excel 2019, then the array will return an error.

If this is the case for you, you will need to link the first cell and drag it down and to the right.

	AF	AG	AH	AI	AJ
35					
36		Period		=VW36	=VX36
37		Start Date		=VW37	=VX37
38		End Date		=VW38	=VX38
39		=Z39			
40		=Z40		=CM40+EQ40+GU40+IY40+LC40	=CN40+ER40+GV40+IZ40+LD40
41		=Z41		=CM41+EQ41+GU41+IY41+LC41	=CN41+ER41+GV41+IZ41+LD41
42		Total		=SUM(AI39:AI41)	=SUM(AJ39:AJ41)

Figure 5.25 Review 6.

CAPEX

In this model, I will assume that the capital expenditure (CAPEX) is not a full retrofit or extensive renovation but mainly a 'light-touch' refurbishment when the tenant leaves. As such, the start of the CAPEX will happen at the lease end date, and we will have an assumption for the refurbishment duration.

Refurbishment Costs

There are two caveats in the use of CAPEX in this instance. First is that I am assuming that the CAPEX does not increase with inflation or other growth rate over time. If you think that will be the case, you can increase the estimates by a growth series, just like we did with the ERV using the future value formula:

FV (CAPEX) = PV (CAPEX) × (1 + growth)^period

Inputs:

	MX	MY	MZ	NA
34				
35	CAPEX			
36				
37				
38				
39	Tenant	Lease End	Refurb End	Capex
40	Tenant 1	27-Sep-30	27-Nov-30	150,000
41	Tenant 2	12-Jun-33	12-Sep-33	180,000
42	Total			330,000

	MX	MY	MZ	NA
34				
35	CAPEX			
36				
37				
38				
39	=D39	Lease End	Refurb End	Capex
40	=D40	=IR40	=EDATE(MY40,refurb duration for T1)	=Capex for T1
41	=D41	=IR41	=EDATE(MY41,refurb duration for T2)	=Capex for T2
42	Total			=SUM(NA40:NA41)

Figure 5.26 Capex.

The other caveat is that the distribution of the CAPEX will follow a straight-line, which means that the distribution will be the same over the duration of the refurbishment. For example, if the total CAPEX is 1 million over 5 months, then each month, it will amount to 200,000 (1 million divided by 5). In our case, we will divide the CAPEX by the YEARFRAC of the refurb duration. The amount will be apportioned according to the DAYS360 function as the standard in our model. We will further see other ways to distribute CAPEX when we do the development model, especially the S-curve.

We will also recognise CAPEX as a negative value (because it is a cost) and in arrears, which means that we will never see a CAPEX amount in period zero, and the references to the cash flow dates will be of the previous period and not the current one. For example, if the refurbishment starts on 31/12/2025, then the first amount will appear in the period between 31/12/2025 and 31/03/2026 and not the period between 30/09/2025 and 31/12/2025.

So our formula for the CAPEX will be:

= - MAX(DAYS360 (MAX(Lease End, <u>Previous</u> Cash Flow Period Start Date),

MIN(Refurb End Date, <u>Previous</u> Cash Flow Period End Date),TRUE),0))/360

x CAPEX/YEARFRAC(Lease End, Refurb End)

I will also add a 'switch', which has the same effect of an IF function, but the 'switch' is a TRUE or FALSE statement that will make the formula 'appear' or 'disappear'. In Excel, if a statement is TRUE, then it will be '1', or if FALSE, it will be '0 (zero)'. If we multiply a number or a formula, in our case, with a TRUE or FALSE statement, then this calculation will turn zero if the statement returns a FALSE.

In our case, we want the formula to 'appear' when the period of the cash flow is greater than zero, because CAPEX will be recognised in arrears. In this case, our 'switch' will be: **(Period >0)**. If the period is greater than zero, then we want the formula for CAPEX to 'appear', and if the Period = 0, then we want the result to be zero. To achieve this, we just need to *multiply* the switch by the formula. It will then be:

= - (Period >0) × MAX (DAYS360...))

	NE	NF	NG	NH	NI	NJ	OG	OH	OI	OJ	OS	OT
34												
35												
36	Period		0	1	2	3	26	27	28	29	38	39
37	Date		28-Feb-24	31-Mar-24	30-Jun-24	30-Sep-24	30-Jun-30	30-Sep-30	31-Dec-30	31-Mar-31	30-Jun-33	30-Sep-33
38	End Date		31-Mar-24	30-Jun-24	30-Sep-24	31-Dec-24	30-Sep-30	31-Dec-30	31-Mar-31	30-Jun-31	30-Sep-33	31-Dec-33
39	**Tenant**											
40	Tenant 1		0	0	0	0	0	-7,500	-142,500	0	0	0
41	Tenant 2		0	0	0	0	0	0	0	0	-36,000	-144,000
42	**Total**		0	0	0	0	0	-7,500	-142,500	0	-36,000	-144,000

	NE	NF	NG
34			
35			
36	Period	=LC36#	
37	Date		
38	End Date		
39	=MX39		
40	=MX40		=-(NG$36>0)*(MAX(DAYS360(MAX($MY40,NF$37),MIN($MZ40,NF$38),TRUE),0))/360*$NA40/YEARFRAC($MY40,$MZ40)
41	=MX41		=-(NG$36>0)*(MAX(DAYS360(MAX($MY41,NF$37),MIN($MZ41,NF$38),TRUE),0))/360*$NA41/YEARFRAC($MY41,$MZ41)
42	=MX42		=SUM(NG40:NG41)

Figure 5.27 Capex 2.

OPEX

Void Costs

Void costs refer to the costs associated with a vacancy. When there is a tenant in place, then the tenant will be responsible for the service charge (remember, we are assuming rents to be taken as net in our model) and any possible repairs, maintenance, insurance, and occupancy taxes, which in the UK they are referred to as 'business rates'. However, if the unit is vacant, then what used to be a tenant's cost will now become the owner's cost. We call this 'void costs'.

In our model, we will have two void periods: an initial one if there is a vacant unit at the time of valuation or acquisition, and another one after the first lease comes to an end through a break or expiry.

Figure 5.28 Void Costs 1.

As with CAPEX, we will assume that the void costs will not increase over time and that costs are recognised in arrears and as negative numbers.

Void 1

	PB	PC	PD
34			
35	**Void 1 - Void Costs**		
36			
37			
38			
39	**Tenant**	**Start**	**Void Costs**
40	Tenant 1	13-Aug-18	75,000
41	Tenant 2	13-Aug-18	90,000
42	**Total**		165,000

P	PB	PC	PD	PE
34				
35	**Void 1 - Void Costs**			
36				
37				
38				
39	=D39		Lease Start	Void Costs
40	=D40		=K40	=V40
41	=D41		=K41	=V41
42	Total			=SUM(PD40:PD41)

Figure 5.29 Void Costs 2.

The formula will be the same DAYS360 as seen before:
=-(Period>0)*(MAX(DAYS360(Previous Period Cash Flow Start Date, MIN(Lease 1 Start Date, Previous Period Cash Flow End Date), TRUE),0))/360 × Void Costs 1
=-(PK$36>0)*(MAX(DAYS360(PJ$37,MIN($PC40,PJ$38), TRUE),0))/360*$PD40

Void 2

R	RF	RG	RH	RI
34				
35	**Void 2 - Void Costs**			
36				
37				
38				
39	**Tenant**	**1 Lease End**	**2 Lease Start**	**Void Costs**
40	Tenant 1	27-Sep-30	27-Jun-31	75,000
41	Tenant 2	12-Jun-33	12-Mar-34	90,000
42	**Total**			165,000

RE	RF	RG	RH	RI
34				
35		**Void 2 - Void Costs**		
36				
37				
38				
39	=PB39	1 Lease End	2 Lease Start	Void Costs
40	=PB40	=CF40	=KU40	=V40
41	=PB41	=CF41	=KU41	=V41

Figure 5.30 Void Costs 3.

Note that I am using the 'arrays' here too, but if you cannot because you are in Mac or in Excel 2019, you will need to link cell by cell as previously explained.

Advanced Discounted Cash Flow Modelling and Analysis 87

The formula will be the same logic (consistency is key in my models as you may have realised by now!):

=-(Period>0)*(MAX(DAYS360(MAX(Lease End, Previous Period Cash Flow Start Date), MIN(Lease 2 Start Date, Previous Period Cash Flow End Date), TRUE),0))/360 × Void Costs 2
=-(RO$36>0)*(MAX(DAYS360(MAX($RG40,RN$37),MIN($RH40,RN$38),TRUE),0))/360*$RI40

Letting Fees

The letting fees are the commission given to the broker for finding a tenant. This commission is typically given as a percentage of the annual rent or ERV at the time of signing the contract.

As such, the letting fee is not apportionable as it is a one-off paid at the signing of the lease contract, at the Lease Start Dates to be more precise.

Figure 5.31 Letting Fees.

The inputs will be Lease 1 Start Date, Lease 2 Start Date, Lease 1 ERV, and Lease 2 ERV. Note that the Lease 1 ERV cannot be taken as the Day 1 ERV because if there is an initial vacancy and the Lease 1 Start Date will be in the future, then the ERV will need to be increased by the growth rate, hence the need to link the ERV with the growth rate series using the LOOKUP function.

In the Input table, we only need to look more closely to the Lease 1 ERV, because all other inputs have been linked from previous calculations.

The Lease 1 ERV calculation will be:
=IF (Lease 1 Start Date < Valuation Date, 0, ERV × LOOKUP(Lease Start 1 Date, Cash Flow Dates, Rental Growth Series)

	TJ	TK	TL	TM	TN
34					
35	Letting Fees				
36					
37					
38					
39	Tenant	Lease 1 Start	Lease 2 Start	Lease 1 ERV	Lease 2 ERV
40	Tenant 1	13-Aug-18	27-Jun-31	0	212,455
41	Tenant 2	13-Aug-18	12-Mar-34	0	269,504
42	Total				

	TI	TJ	TK	TL	TM	TN
34						
35		Letting Fees				
36						
37						
38						
39		=D39	Lease 1 Start	Lease 2 Start	Lease 1 ERV	Lease 2 ERV
40		=D40	=PC40	=KU40	=IF(TK40<$D$10,0,I40*LOOKUP(TK40,$CM$37:$EE$37,$CP$34:$EE$34))	=KW40
41		=D41	=PC41	=KU41	=IF(TK41<$D$10,0,I41*LOOKUP(TK41,$CM$37:$EE$37,$CP$34:$EE$34))	=KW41

Figure 5.32 Letting Fees 2.

Now, because we are not apportioning the letting fees because they are one-off fees, we can start thinking of it as a series of IF functions because if the lease start is between the start and end dates of that period, then it means that the letting fee will be due.

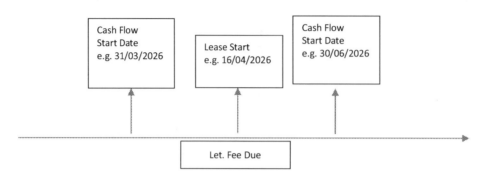

Figure 5.33 Letting Fees 3.

= IF (AND (Cash Flow Start Date < Lease 1 Start, Cash Flow End Date >= Lease 1 Start), ERV 1 x Let Fee, 0)

Because we now know that Excel uses TRUE and FALSE as 1 and 0, we can replace the AND by a multiplication:

= IF ((Cash Flow Start Date < Lease 1 Start) **x** (Cash Flow End Date >= Lease 1 Start), ERV 1 x Let Fee, 0)

We can take this multiplication even further and get rid of the IF function:

= (Cash Flow Start Date < Lease 1 Start) **x** (Cash Flow End Date >= Lease 1 Start) **x** ERV 1 x Let Fee

And then, potentially, we can add the second lease start dates by adding the same if functions but changing Lease 1 to Lease 2 Start and ERV 1 to ERV 2.

However, Excel has a function called SUMPRODUCT, which as the name suggests adds multiplications. So, instead of adding the new multiplication to the formula, we will use the SUMPRODUCT instead:

= -SUMPRODUCT ((Cash Flow Start Date < Lease 1:Lease2) **x** (Cash Flow End Date >= Lease 1:Lease2) **x** (ERV 1:ERV 2) x Let Fee)

Note that the dates need to be part of one array, and the ERVs part of another array, so that the TRUE or FALSE of Start 1 Date can be multiplied by the ERV 1, and the Start 2 Date can be multiplied by the ERV 2.

T	TJ	TK	TL	TM	TN	TCT	TQ	TR	TS
34									
35	Letting Fees								
36							Period		0
37							Date		28-Feb-24
38							End Date		31-Mar-24
39	Tenant	Lease 1 Start	Lease 2 Start	Lease 1 ERV	Lease 2 ERV		Tenant		
40	Tenant 1	13-Aug-18	27-Jun-31	0	212,455		Tenant 1		=-SUMPRODUCT(($TS$37<($TK40:$TL40))*($TS$38>=($TK40:$TL40))*($TM40:$TN40)*$D$25)
41	Tenant 2	13-Aug-18	12-Mar-34	0	269,504		Tenant 2		0
42	Total						Total		0

Figure 5.34 Letting Fees 4.

Review Fees

Review fees follow the same logic as the letting fees, as it is a percentage of an annual ERV and paid as one-off. As such, we will continue to use the SUMPRODUCT.

In our model, we have three review dates, but if you have more, adding them would be easy by expanding the arrays. Just remember that you will need to keep the dates in adjacent cells and the ERVs in a similar fashion so to form the 'arrays'.

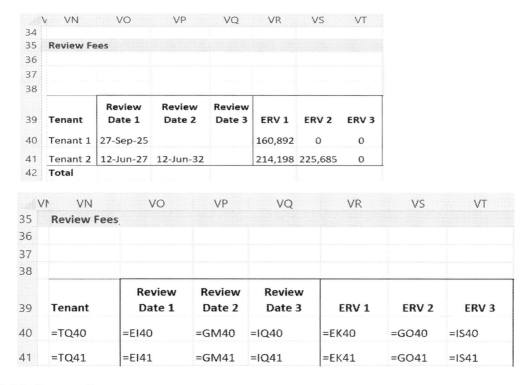

Figure 5.35 Review Fees.

The formula of the review fees will be:
= - SUMPRODUCT((Cash Flow Start Date <Review 1: Review 3) x
(Cash Flow End Date >=Review 1: Review 3)*(ERV 1: ERV 3) x Review Fee))
=-SUMPRODUCT((VW$37<($VO40:$VQ40))*(VW$38>=($VO40:$VQ40))*($VR40:$VT40)*$D$26)

Pro-Forma Cash Flow

Now that all intermediate calculations have been accomplished, we can move to putting everything together in the pro-forma cash flow (or pro-forma in short). This will show the expected cash flow that will be generated by the property. The cash flow is split into three: cash flow from operations (which will give us the net operating income – or NOI), the cash flow from investment (which will give us the capital invested and returned), and the cash flow from financing.

In the next section, we will discuss and model the cash flow from operations and investment, and the net cash flow (NCF) result will be assumed as 100% equity, so they will either be equity calls (if the NCF is negative) or dividends and capital return (if positive). We will further discuss the cash flow from financing when we model the debt component of the investment in the next chapter. As such:

Net Cash Flow = NOI + Cash Flow from Investment

Net Operating Income (NOI)

The NOI is the Operating Revenue (Rents) – Operating Expenditures (OpEx).

Sometimes, the NOI can be considered the EBITDA (earnings before interest, tax, depreciation, and amortisation). My only problem with EBITDA – hence still calling it NCF – is that earnings, if taken from an accounting perspective, may include accruals, for example, instead of considering a rent-free period, the rent may be spread over the life of the lease at a lower rate and the rent-free period not shown in the cash flow as zero. This can have certain consequences for debt covenant ratios and general cash management when the forecast is used in the asset management phase.

In the case of rents being paid net, as in our case, the NOI will be the same as the net rent when the property is stabilised – i.e., full occupancy and all tenants paying normal rents (no rent-frees or other abatements).

Net rents tend to be a term used more in the UK and Continental Europe, and NOI seems to be used more in the US or when rents are received gross and then the owner needs to deduct the operating costs to get to the NOI.

Revenue

In our case, the revenue is the gross rents received from the tenants, and it is just the line below:

Net Rents

= Total of Rents in the Rent Forecast Table below the Pro-Forma

Note: DO NOT just add Rent Tenant 1 + Rent Tenant 2, either link to the Total Cell or use the SUM() function. Why? Because as we need to add new tenants in the model, if you just add Tenant 1 + Tenant 2, the new tenants will need to be added manually which will be a source of error.

Operating Expenses

Operating expenses are the Void Costs + Letting Fees + Review Fees as we are assuming that rents are net.

Void Costs = Void Costs 1 + Void Costs 2 from the Intermediate Calculation
Letting Fees = From the Letting Fees Intermediate Calculation
Review Fees = From the Review Fees Intermediate Calculation

Net Operating Income

The net operating income (NOI) is the revenue + operating expenses (OPEX).

Investment Cash Flow

The investment cash flow here will be purchase costs, CAPEX, and sale price.

Purchase Price

Purchase Price = - Net Value

Because my assumption is that this transaction happens on day 1 and there is no option or deposit paid in advance, I will just leave the cells and formulas as they are and will not copy it to the right.

Purchaser's Costs

Purchaser's Costs = Purchase Price × Purchaser's Costs %

Refurbishment CAPEX

= CAPEX from Calculation Table

Sale Price/Exit Value

You will need to add the cash flow date for the exit date in the Intermediate Calculation Table, using the formula:
The sale price will be received at the end of the holding period or exit date:
= (Period = Holding Period × 4) × Sale Price
Sale Costs
= - Sale Price × Sale Costs %

Net Cash Flow

The net cash flow (NCF) is the money that is left after operating expenses and capital expenditures.
The NCF is thus:
= NOI + CFI (Cash Flow from Investment)

Holding Period

To make the cash flow flexible to changes in the holding period, we can add a technique called 'switch' to the cash flow. Switch means making things appear or disappear in Excel.

In this case, the switch will be:
= (Period < Holding Period x 4) x FORMULA
FORMULA is what is there already, and if the cash flow date is AFTER the exit date, it means that the cell value will be zero (0) no matter the result of the formula.

To be absolutely robust, we should add this 'switch' in all formulas of the cash flow, but in the interest of time, we will add it only to headline formulas, i.e., revenue, operating expenses, and investment cash flow.

Revenue Holding Period Switch

= (Period < Holding Period × 4)

Note that this is only < and not <=, because we are assuming rents are received in advance. To be consistent, if you assume to be receiving rents in Period 0, you should not be receiving rent in the last period.

Operating Expenses and Investment Cash Flow

= (Period<= Holding Period × 4)

Here, we will need to use <= because we are assuming costs paid in arrears, so any costs in the last period of the holding period should be accounted for.

	AI	AJ	AK	AL	AM	AN	BI	BJ	BK
Pro-forma Cash Flow									
Period	0	1	2	3	4	5	26	27	28
Start Date	28-Feb-24	31-Mar-24	30-Jun-24	30-Sep-24	31-Dec-24	31-Mar-25	30-Jun-30	30-Sep-30	31-Dec-30
End Date	31-Mar-24	30-Jun-24	30-Sep-24	31-Dec-24	31-Mar-25	30-Jun-25	30-Sep-30	31-Dec-30	31-Mar-31
Revenue	29,867	84,000	84,000	84,000	84,000	84,000	87,269	49,684	49,684
Net Rent	29,867	84,000	84,000	84,000	84,000	84,000	87,269	49,684	49,684
Operating Expenses (OPEX)	-	-	-	-	-	-	-	(625)	(18,750)
Void Costs	-	-	-	-	-	-	-	(625)	(18,750)
Letting Fees	-	-	-	-	-	-	-	-	-
Review Fees	-	-	-	-	-	-	-	-	-
Net Operating Income (NOI)	29,867	84,000	84,000	84,000	84,000	84,000	87,269	49,059	30,934
Investment Cash Flow	(6,109,091)	-	-	-	-	-	-	(7,500)	(142,500)
Purchase Price	(5,720,123)								
Purchaser's Costs	(388,968)								
Refurb CAPEX	-	-	-	-	-	-	-	(7,500)	(142,500)
Sale Price	-	-	-	-	-	-	-	-	-
Sale Costs	-	-	-	-	-	-	-	-	-
Net Cash Flow (NCF)	(6,079,224)	84,000	84,000	84,000	84,000	84,000	87,269	41,559	(111,566)

Figure 5.36 Holding Period 1.

	AH	AI	AJ
Pro-forma Cash Flow			
=AG36:AG38		=AI36#	
Revenue		=(AI4<$D$12*4)*AI8	=(AJ4<$D$12*4)*AJ8
Net Rent		=AI42	=AJ42
Operating Expenses (OPEX)		=(AI4 <= $D$12*4)* SUM(AI11:AI13)	=(AJ4 <= $D$12*4)* SUM(AJ11:AJ13)
Void Costs		=PK42+RO42	=PL42+RP42
Letting Fees		=TS42	=TT42
Review Fees		=VW42	=VX42
Net Operating Income (NOI)		=AI7+AI10	=AJ7+AJ10
Investment Cash Flow		=(AI4 <= $D$12*4) * SUM(AI18:AI22)	=(AJ4 <= $D$12*4) * SUM(AJ18:AJ22)
Purchase Price		=-D23	
Purchaser's Costs		=AI18*D27	
Refurb CAPEX		=NG42	=NH42
Sale Price		=(AI4=$D$12*4)*$E$23	=(AJ4=$D$12*4)*$E$23
Sale Costs		=-AI21*$D$28	=-AJ21*$D$28
Net Cash Flow (NCF)		=AI15+AI17	=AJ15+AJ17

Figure 5.37 Holding Period 2.

Returns Calculation

Now that you have calculated the whole cash flow, it is time to check if it is a good investment or not.

We will calculate the IRR, NPV, worth, profit, equity, and equity multiple.

Note that whilst we previously used the function =**IRR** and =**NPV**, we will now use the functions =**XIRR** and =**NPV**.

IRR vs XIRR

The difference between IRR and XIRR is that the IRR will assume cash flows are annual, whereas the XIRR can calculate the IRR even if the cash flows do not follow the same frequency, for example, one cash flow is monthly and the other semi-annual; so, the distance between cash flows can be different. For the XIRR to work though, you will need to tell Excel the dates of each cash flow, so then Excel will calculate a daily rate and annualise it. As such, the result of the XIRR is already an annual rate and you will not need to annualise it.

The syntax for XIRR is thus:

= **XIRR** (cash flow range, cash flow date range)

NPV vs XNPV

The idea is the same for the XNPV, the cash flows do not need to have the same time distance from each other, and we will need the cash flow dates for each cash flow. The good news is that Excel will know what to discount or not. Remember that in NPV, Excel assumes that the cash flows are in arrears, and it discounts the first value. Because our cash flows are in advance (we start in Period 0 and assume it is today), we had to exclude the Period 0 (zero) cash flow and add it back at the end.

Another good news is that we won't need to de-compound the discount rate according to the periodicity of the cash flow. With NPV, we need to calculate a monthly discount rate for a monthly cash flow, quarterly rate for a quarterly cash flow, and so on. Now, with XNPV, we can use an annual discount rate.

The syntax for XNPV is:

=XNPV (annual discount rate, cash flow range, cash flow dates)

The only times that the XIRR and XNPV will not work is when there is no cash flow in the first period or when this number is very small in relation to the whole cash flow. In those cases, you will need to revert to IRR and NPV calculations.

Price vs Value vs Worth

The way I see it is that price is the offering price in the market when being sold or price at which the property was transacted, so it really refers to something more concrete than value. Value is then the hypothetical value, a market value that a valuer has estimated. Worth is then the investment value, how much a particular investor would pay for an asset. Each investor will have their own idea of worth, since they have different assumptions and views for the future. The worth for an investor would therefore be the maximum bidding value for the property.

If you are doing a market valuation, then the worth and the value will be the same, so you can start with the initial value as zero and then work backwards to find out the market value as worth.

Profit

Profit is the SUM of the NCFs, so it's the amount left after all the outflows and inflows throughout the cash flow.

Equity/Peak Equity

First, we considered as equity only the equity used to purchase the property. However, in the investment world, people talk about 'peak equity', which includes any negative cash flows after the acquisition of the asset. For example, in periods when there may be a vacancy and void costs or CAPEX, it is assumed that this amount will come as equity since we are not considering any debt or leverage.

To calculate this, we would need the **SUMIF** function, so that we sum all the negative cash flows (values < 0). The syntax of SUMIF is:

= **SUMIF** (range, criteria)
= **SUMIF** (cash flow range, "<0")

Note that for the negative criteria, we will need **"<0"**.

Return on Equity (ROE) and Equity Multiple

The return on equity (ROE) is the profitability margin of the investment:

ROE = Profit/Equity

The equity multiple is how many times your one pound or dollar will multiply over the life of the investment. Thus, if your profit margin (ROE) is 50%, that means that for each pound you put in, you will make 50 pence in profit. In terms of equity multiple, it means that for each pound that you put in, you will receive back 1.5x. As such, the equity multiple formula is:

Equity Multiple = Profit/Equity + 1
Equity Multiple = ROE + 1

Output Table/Returns Calculation

	G	H	I	J
7				
8		OUTPUT		
10		KPIs		
11		IRR		7.74%
12		NPV		106,142
13		Net Worth (IV)		5,819,506
14		Profit		5,099,356
15		Equity		6,306,031
16		ROE		80.86%
17		Equity Multiple		1.81x

	G	H	I	J
7				
8		OUTPUT		
10		KPIs		
11		IRR		=XIRR(AI24:CA24,AI5:CA5)
12		NPV		=XNPV(D13,AI24:CA24,AI5:CA5)
13		Net Worth (IV)		=(J12+D22)/(1+D27)
14		Profit		=SUM(AI24:CA24)
15		Equity		=-SUMIF(AI24:CA24,"<0")
16		ROE		=J14/J15
17		Equity Multiple		=J16+1

Figure 5.38 Returns.

CASE STUDY 3

London SW1, Office Prime West End Investment

Investment Summary

- Freehold
- Prime location in Mayfair, London

- Refurbished in 2018 to the highest of standards
- 61,381 sq ft (5,702.5 sq m) of Grade A offices
- Total income of £7,308,024 per annum (average office rent of £118 per sq ft)

The tenancy schedule is:

Demise	Tenant	Floor Area (sq ft)	Term Start	Term Expiry	Next Review	Break Date	Rent £ pa	Rent £ PSF
Sixth Floor	Tenant 1	6,395	27 Sep 2019	26 Sep 2034	27 Sep 2024	27 Sep 2031	901,496	140
Fifth Floor	Tenant 2	8,240	27 Sep 2019	26 Sep 2034	27 Sep 2024	27 Sep 2031	994,996	120
Fourth Floor	Tenant 3	9,228	27 Sep 2019	26 Sep 2034	27 Sep 2024	27 Sep 2031	1,113,556	120
Third Floor	Tenant 4	9,784	27 Sep 2019	26 Sep 2034	27 Sep 2024	27 Sep 2031	1,174,080	120
Second Floor	Tenant 5	8,135	27 Sep 2019	26 Sep 2034	27 Sep 2024	27 Sep 2031	756,555	93
First Floor	Tenant 6	9,496	27 Sep 2019	26 Sep 2034	27 Sep 2024	27 Sep 2031	1,240,676	130
Ground Floor	Tenant 7	7,127	18 Jul 2019	17 Jul 2029	18 Jul 2024	17 Jul 2024	862,240	120
Lower Ground	Tenant 8	2,976	18 Jul 2019	17 Jul 2029	18 Jul 2024	17 Jul 2024	264,424	86.5
Total		**61,381**					**7,308,024**	

Figure 5.39 Multi-Let Case Study Tenancy Schedule.

To use the model we have just created, all you need to do is add the tenancy schedule to the model and find comparable evidence to make further estimates. To add new tenants, you will need to add rows in the tenancy schedule, and you should add *between* Tenants 1 and 2. The reason for that is that all ranges will then expand accordingly, and you will not need to adjust any formulas, but only copy down.

You should highlight the entire row for Tenant 2 and add rows, by right clicking or pressing CTRL +.

96 Real Estate Financial Modelling in Excel

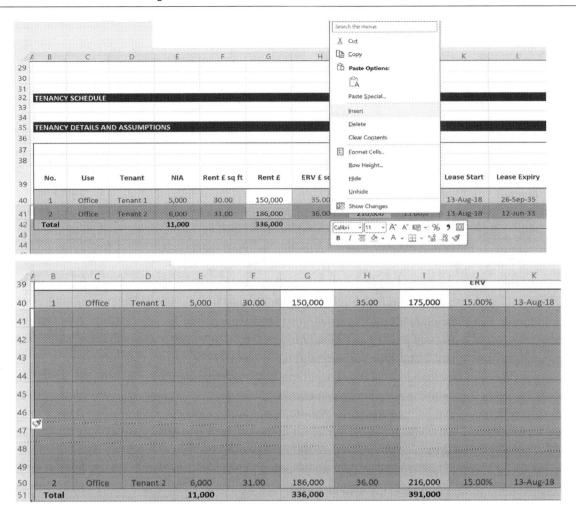

Figure 5.40 Multi-Let Case Study Adding Rows.

Once you have done that, you can do the data entry.

No.	Use	Tenant	NIA	Rent £ sq ft	Rent £	ERV £ sq ft	ERV £	Discount to ERV	Lease Start	Lease Expiry	Next Review	Review Cycle
6th	Office	Tenant 1	6,395	141	901,496	155.07	991,646	15.00%	27-Sep-19	26-Sep-34	27-Sep-24	5 years
5th	Office	Tenant 2	8,240	121	994,996	132.83	1,094,496	15.00%	27-Sep-19	26-Sep-34	27-Sep-24	5 years
4th	Office	Tenant 3	9,228	121	1,113,556	132.74	1,224,912	15.00%	27-Sep-19	26-Sep-34	27-Sep-24	5 years
3rd	Office	Tenant 4	9,784	120	1,174,080	132.00	1,291,488	15.00%	27-Sep-19	26-Sep-34	27-Sep-24	5 years
2nd	Office	Tenant 5	8,135	93	756,555	102.30	832,211	15.00%	27-Sep-19	26-Sep-34	27-Sep-24	5 years
1st	Office	Tenant 6	9,496	131	1,240,676	143.72	1,364,744	15.00%	27-Sep-19	26-Sep-34	27-Sep-24	5 years
G	Office	Tenant 7	7,127	121	862,240	133.08	948,464	15.00%	18-Jul-19	17-Jul-34	18-Jul-24	5 years
LG	Office	Tenant 8	2,976	89	264,424	97.74	290,866	15.00%	18-Jul-19	17-Jul-34	18-Jul-24	5 years
Total			61,381	119.06	7,308,023	130.97	8,038,825					

No.	Use	Upwards Only?	Break Date	Assume Break	Refurb Costs £ sq ft	Total Refurb	Refurb Duration	Void Costs sq ft	Total	Void	Rent Free
6th	Office	N	27-Sep-31	N	80.00	511,600	3 months	50.00	319,750	12 months	24 months
5th	Office	N	27-Sep-31	N	80.00	659,200	3 months	50.00	412,000	12 months	24 months
4th	Office	N	27-Sep-31	N	80.00	738,240	3 months	50.00	461,400	12 months	24 months
3rd	Office	N	27-Sep-31	N	80.00	782,720	3 months	50.00	489,200	12 months	24 months
2nd	Office	N	27-Sep-31	N	80.00	650,800	3 months	50.00	406,750	12 months	24 months
1st	Office	N	27-Sep-31	N	80.00	759,680	3 months	50.00	474,800	12 months	24 months
G	Office	N	17-Jul-24	N	80.00	570,160	3 months	50.00	356,350	12 months	24 months
LG	Office	N	17-Jul-24	N	80.00	238,080	3 months	50.00	148,800	12 months	24 months
Total									3,069,050		

Figure 5.41 Multi-Let Case Study Tenancy Schedule 2.

Note that the assumptions are my own based on what I know today and the types of comparables I found. Once the inputs have been changed to the reflect the reality of each property, its tenants and tenancy agreements, the model will calculate itself.

The Input and Output tables as shown as figure.

GLOBAL INPUTS		
Date	28-Feb-24	
Holding Period	10 yrs	
Target Rate	7.00%	
	Entry	Exit
Risk-Free Rate	4.00%	3.00%
Risk Premium	3.00%	3.00%
Rental Growth	3.00%	2.00%
Net Initial Yield	4.00%	4.00%
Total Rent	7,308,023	10,803,509
Gross Value	182,700,575	270,087,725
Net Value	171,067,954	270,087,725
Letting Fee	15.00%	
Review Fee	10.00%	
Purchaser's Cost	6.80%	
Sale Cost	0.50%	

OUTPUT	
KPIs	
IRR	7.59%
NPV	8,615,618
Net Worth (IV)	179,135,012
Profit	159,482,938
Equity	182,050,973
ROE	87.60%
Equity Multiple	1.88x

GROWTH AND DEPRECIATION ASSUMPTIONS

Year | Rental Growth

Year	Market Growth	Depreciation	Rental Growth
2024	4.00%	1.00%	3.00%
2025	4.00%	1.00%	3.00%
2026	4.00%	1.00%	3.00%
2027	4.00%	1.00%	3.00%
2028	4.00%	1.00%	3.00%

Figure 5.42 Multi-Let Case Study Summary.

Pro-forma Cash Flow

	AI	AJ	AK	AL	AM	AN	AO	AP	BV	BW
Period	0	1	2	3	4	5	6	7	39	40
Start Date	*28-Feb-24*	*31-Mar-24*	*30-Jun-24*	*30-Sep-24*	*31-Dec-24*	*31-Mar-25*	*30-Jun-25*	*30-Sep-25*	*30-Sep-33*	*31-Dec-33*
End Date	31-Mar-24	30-Jun-24	30-Sep-24	31-Dec-24	31-Mar-25	30-Jun-25	30-Sep-25	31-Dec-25	31-Dec-33	31-Mar-34
Revenue	649,602	1,827,006	1,815,498	1,751,063	1,751,063	1,751,063	1,751,063	1,751,063	1,942,910	-
Net Rent	649,602	1,827,006	1,815,498	1,751,063	1,751,063	1,751,063	1,751,063	1,751,063	1,942,910	1,942,910
Operating Expenses (OPEX)	-	-	(700,425)	-	-	-	-	-	-	-
Void Costs	-	-	-	-	-	-	-	-	-	-
Letting Fees	-	-	-	-	-	-	-	-	-	-
Review Fees	-	-	(700,425)	-	-	-	-	-	-	-
Net Operating Income (NOI)	649,602	1,827,006	1,115,073	1,751,063	1,751,063	1,751,063	1,751,063	1,751,063	1,942,910	-
Investment Cash Flow	(182,700,575)	-	-	-	-	-	-	-	-	268,737,286
Purchase Price	(171,067,954)									
Purchaser's Costs	(11,632,621)									
Refurb CAPEX	-	-	-	-	-	-	-	-	-	-
Sale Price	-	-	-	-	-	-	-	-	-	270,087,725
Sale Costs	-	-	-	-	-	-	-	-	-	(1,350,439)
Net Cash Flow (NCF)	(182,050,973)	1,827,006	1,115,073	1,751,063	1,751,063	1,751,063	1,751,063	1,751,063	1,942,910	268,737,286

Figure 5.43 Multi-Let Case Study Cash Flow.

What Is the Market Value?

There are two options here. You can either assume that the NIY is right, i.e., 4.0%, and in this case, all your evidence is based on 'market' instead of a particular investor's assumption, the net value is the market value, i.e., 171,067,954 and that will give you an IRR of 7.59%.

On the other hand, you can assume that the discount rate of 7.0% is the correct benchmark and then in this case the market value will be the net worth or 179,135,012. This can be demonstrated if you run a Goal Seek or Solver by setting the discount rate as a hardcoded value of 7.0% (you cannot link it to the risk-free + risk premium anymore; otherwise, it will be a circular reference) and setting the NPV equal to zero and changing the risk premium. You can also change the risk-free rate or the rental growth, but my preference is for changing the risk premium as this can be seen as a less observable value than the risk-free or rental growth and easier to justify as a market valuation. The net initial yield then becomes 3.82% instead of 4.0%.

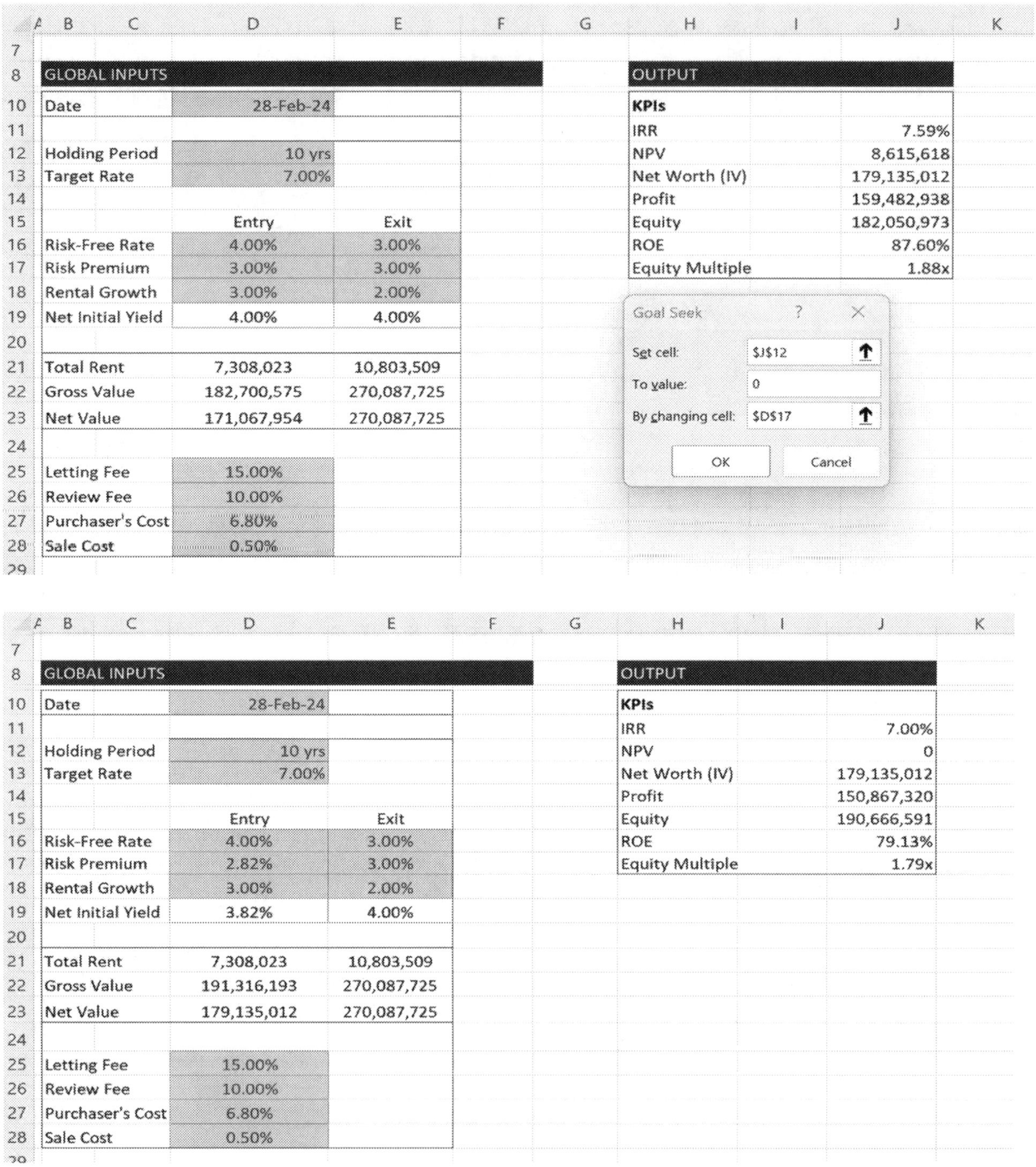

Figure 5.44 Multi-Let Case Study 1.

I would argue that the 'correct' way to do a market valuation would be based on the discount rate rather than the initial yield, particularly when a property is not stabilised or 'rack rented'. Also, the discount rate can be compared not only within real estate assets but also across other financial assets such as corporate bonds and equities.

Advanced Discounted Cash Flow Modelling and Analysis

Creating a Summarised Annual Cash Flow

An annual cash flow should be created only once a more granular cash flow has been modelled.

I will create a separate sheet for the annual cash flow and link it to the quarterly cash flow sheet, which I named Qtr_CF.

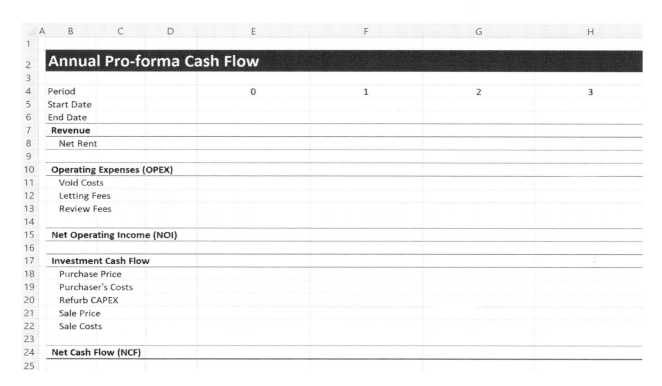

Figure 5.45 Multi-Let Case Study 2.

Figure 5.46 Multi-Let Case Study 3.

100 Real Estate Financial Modelling in Excel

	A	B	C	D	E
1					
2		**Annual Pro-forma Cash Flow**			
3					
4		=Qtr_CF!Z4			0
5		=Qtr_CF!Z5			=Qtr_CF!AI5
6		=Qtr_CF!Z6			=Qtr_CF!AI6
7		**=Qtr_CF!Z7**			=SUMPRODUCT((E$5<=Qtr_CF!$AI$5:$CA$5)*(E$6>=Qtr_CF!$AI$6:$CA$6)*(Qtr_CF!$AI7:$CA7))
8		=Qtr_CF!Z8			=SUMPRODUCT((E$5<=Qtr_CF!$AI$5:$CA$5)*(E$6>=Qtr_CF!$AI$6:$CA$6)*(Qtr_CF!$AI8:$CA8))
9					
10		**=Qtr_CF!Z10**			=SUMPRODUCT((E$5<=Qtr_CF!$AI$5:$CA$5)*(E$6>=Qtr_CF!$AI$6:$CA$6)*(Qtr_CF!$AI10:$CA10))
11		=Qtr_CF!Z11			=SUMPRODUCT((E$5<=Qtr_CF!$AI$5:$CA$5)*(E$6>=Qtr_CF!$AI$6:$CA$6)*(Qtr_CF!$AI11:$CA11))
12		=Qtr_CF!Z12			=SUMPRODUCT((E$5<=Qtr_CF!$AI$5:$CA$5)*(E$6>=Qtr_CF!$AI$6:$CA$6)*(Qtr_CF!$AI12:$CA12))
13		=Qtr_CF!Z13			=SUMPRODUCT((E$5<=Qtr_CF!$AI$5:$CA$5)*(E$6>=Qtr_CF!$AI$6:$CA$6)*(Qtr_CF!$AI13:$CA13))
14					
15		**=Qtr_CF!Z15**			=SUMPRODUCT((E$5<=Qtr_CF!$AI$5:$CA$5)*(E$6>=Qtr_CF!$AI$6:$CA$6)*(Qtr_CF!$AI15:$CA15))
16					
17		**=Qtr_CF!Z17**			=SUMPRODUCT((E$5<=Qtr_CF!$AI$5:$CA$5)*(E$6>=Qtr_CF!$AI$6:$CA$6)*(Qtr_CF!$AI17:$CA17))
18		=Qtr_CF!Z18			=SUMPRODUCT((E$5<=Qtr_CF!$AI$5:$CA$5)*(E$6>=Qtr_CF!$AI$6:$CA$6)*(Qtr_CF!$AI18:$CA18))
19		=Qtr_CF!Z19			=SUMPRODUCT((E$5<=Qtr_CF!$AI$5:$CA$5)*(E$6>=Qtr_CF!$AI$6:$CA$6)*(Qtr_CF!$AI19:$CA19))
20		=Qtr_CF!Z20			=SUMPRODUCT((E$5<=Qtr_CF!$AI$5:$CA$5)*(E$6>=Qtr_CF!$AI$6:$CA$6)*(Qtr_CF!$AI20:$CA20))
21		=Qtr_CF!Z21			=SUMPRODUCT((E$5<=Qtr_CF!$AI$5:$CA$5)*(E$6>=Qtr_CF!$AI$6:$CA$6)*(Qtr_CF!$AI21:$CA21))
22		=Qtr_CF!Z22			=SUMPRODUCT((E$5<=Qtr_CF!$AI$5:$CA$5)*(E$6>=Qtr_CF!$AI$6:$CA$6)*(Qtr_CF!$AI22:$CA22))
23					
24		**=Qtr_CF!Z24**			=SUMPRODUCT((E$5<=Qtr_CF!$AI$5:$CA$5)*(E$6>=Qtr_CF!$AI$6:$CA$6)*(Qtr_CF!$AI24:$CA24))

Figure 5.47 Multi-Let Case Study 4.

	D	E	F	G	H	I	J	K	L	M	N	O
2	**Annual Pro-forma Cash Flow**											
4	Period	0	1	2	3	4	5	6	7	8	9	10
5	Start Date	28-Feb-24	31-Mar-24	31-Mar-25	31-Mar-26	31-Mar-27	31-Mar-28	31-Mar-29	31-Mar-30	31-Mar-31	31-Mar-32	31-Mar-33
6	End Date	31-Mar-24	31-Mar-25	31-Mar-26	31-Mar-27	31-Mar-28	31-Mar-29	31-Mar-30	31-Mar-31	31-Mar-32	31-Mar-33	31-Mar-34
7	Revenue	649,602	7,190,856	7,090,193	7,090,193	7,090,193	7,090,193	7,730,406	8,280,441	8,280,441	8,280,441	6,210,331
8	Net Rent	649,602	7,190,856	7,090,193	7,090,193	7,090,193	7,090,193	7,730,406	8,280,441	8,280,441	8,280,441	8,280,441
10	**Operating Expenses (OPEX)**	-	(709,019)	-	-	-	-	(828,044)	-	-	-	-
11	Void Costs											
12	Letting Fees											
13	Review Fees	-	(709,019)	-	-	-	-	(828,044)	-	-	-	-
15	**Net Operating Income (NOI)**	649,602	6,481,837	7,090,193	7,090,193	7,090,193	7,090,193	6,902,362	8,280,441	8,280,441	8,280,441	6,210,331
17	**Investment Cash Flow**	(191,316,193)										268,737,286
18	Purchase Price	(179,135,012)										
19	Purchaser's Costs	(12,181,181)										
20	Refurb CAPEX											
21	Sale Price											270,087,725
22	Sale Costs											(1,350,439)
24	**Net Cash Flow (NCF)**	(190,666,591)	6,481,837	7,090,193	7,090,193	7,090,193	7,090,193	6,902,362	8,280,441	8,280,441	8,280,441	274,947,617

Figure 5.48 Multi-Let Case Study 5.

Then, link the dates. For Period 0, it will be the same start and end dates, and for period 1 onwards, it will be =EDATE (previous end date, 12). Now, all you need to do is create a SUMPRODUCT formula linking the start and end dates of the annual cash flow with the quarterly cash flow:

=SUMPRODUCT ((Annual Start Date <= Qtr Start Date Range) × (Annual End Date >=Qtr End Date Range) × (Cash Flow Item Range))

Or:

=SUMPRODUCT((E$5<=Qtr_CF!$AI$5:$CA$5)*(E$6>=Qtr_CF!$AI$6:$CA$6)*(Qtr_CF!$AI7:$CA7))

Just pay attention to the dollar signs and drag it all down.

Conclusion

In this chapter, we learnt how to create a multi-let cash flow from scratch using a basic template. We started discussing the difficulties in assimilating all lease events in a cash flow, which we called the dating problem. We looked at a UK property, in which reviews take place to market rent, in property jargon, ERV, OMR, or OMRV.

We studied how to incorporate the following lease events into our multi-tenant cash flow: lease start, multiple reviews, lease end, void period, new lease start, rent-free period, and rent forecast using a multiple growth series.

In the next chapter, we will further develop the pro-forma cash flow incorporating debt finance (gearing).

In this chapter, we mastered how to create pro-forma cash flow, including operating expenses, such as void costs and letting fees, investment cash flow including capital expenditures, and purchase and sale costs. We have included a holding period switch to make the cash flow flexible depending on how long the asset would be held for.

As real estate is unique and transactions are always different from one another, with different tenants, exit value assumptions, and tenancy schedules, financial models for real estate investment transactions are unique as well. As such, following this book, you should be able to fine-tune your financial models to incorporate different aspects of each transaction.

Section 2

Financial Modelling of Commercial Real Estate Debt – Gearing (or Leveraging)

Chapter 6

Debt Structures

Chapter Contents

Introduction	105
Debt Structures Description and Framework	106
Debt Structure Types	106
Interest Rate Type	106
Amortisation	110
Amortisation Period	110
Maturity	110
Fees	110
Debt Covenants	111
Input Table	113
Debt Structures Modelling	116
Interest Only (I/O)	118
Formulas	120
Covenants	121
Have Debt Covenants Been Breached?	122
Has Leverage Improved Financial Returns?	123
Outputs Calculations	124
Interest-Only Structure – Complete Cash Flow Formulas	125
Amortising Loans	125
Fully Amortising – Constant Amortisation	126
Partially Amortising – Constant Amount	126
Partially Amortising – Constant Percentage (CA)	127
Constant Amortisation Debt Structure Modelling	128
Constant Amortisation (%) Formula	130
Constant Amortisation – Constant Amount Debt Cash Flow	131
Constant Payment – Full Amortisation (CP)	131
Annuity Calculation – PMT	133
Constant Payment – Balloon (CP-B)	134
Constant Payment with Balloon Debt Structure Modelling	135
Constant Payment with Balloon Amortisation Formula	136
Rolled-Up Interest	136

Rolled-Up Interest Debt Structure Modelling	137
Rolled-Up Interest Payment Formula	138
Cash Sweep	138
Cash Sweep Debt Structure Modelling	139
Cash Sweep Interest Payment Formula (MIN and MAX Functions)	140
Cash Sweep Amortisation Formula	140
Floating Interest Rates	141
The Yield Curve	142
Bank Rate	143
Short-Term Interest Rates	143
Interest Rates Forecast	144
The Forward Curve	144
Covenants and Margins	145
Floating Interest Rates Debt Structure Modelling	145
Conclusion	148

Introduction

Now that you have modelled the asset side of the property investment, we will look at financial engineering this investment, i.e., obtaining debt financing.

In this second part of the book, we will look at *commercial real estate debt* as part of the capital structure for the investment. As such, debt in our models are taken from a borrower's perspective, meaning that interest payments, amortisation, and final capital repayments are cash outflows and represented as negative values. If you are, however, a lender, you will need to swap the signs around so that interest, amortisation, and final capital repayments are cash inflows and will have a positive sign in your model. This is mainly a matter of direction of signage, but it is important to note that the fundamentals will remain the same.

We will first model the most common real estate debt repayments structures, namely: interest only, constant amortisation, constant payment, constant payment with balloon, rolled-up, or capitalised interest. We will also add a cash sweep structure, but mainly for you to understand the mechanism and how it is modelling, but sweep structures as loan terms are not commonly practiced, unless loans are in default (which will be a topic for another book!).

Firstly, we will model all structures assuming a fixed rate and then we will add the complication of floating rates based on forward curves. We will then create a mechanism to vary the types of interest and structures and hopefully the model will give you enough insight into what kind of structure you should choose from and also if you should hedge or not hedge your interest payments.

As in modelling real estate assets from an equity perspective, modelling real estate debt is also particular to each agreement, since lenders will have differing terms and conditions of lending.

In the context of real estate investment, 'debt structures' refer to the various ways in which the repayment details of a loan are described, and in this chapter, we will outline the main repayment types, covenants and how to find out the maximum loan amount.

Debt Structures Description and Framework

We will first create a sheet with all the debt types or structures that we will be modelling in this book.

Input Headers for Debt Structures

A	B	C	D	E	F	G	H	I
2	**DEBT DETAILS**							
3								
4	*Debt No.*		1	2	3	4	5	6
5	Type		I/O	CA	CP	CP-B	R-Up	SWp
6	Interest rate type							
7	LTV							
8	Fixed - Int. Rate							
9	Floating - Margin							
10	Amortisation							
11	Amort. Period							
12	Maturity							
13								
14	**Fees**							
15	Origination fee							
16	Rolled fee?							
17								
18	**Covenants**							
19	ICR							
20	DSCR							
21	LTV							
22	Debt Yield							

Figure 6.1 Input Headers.

Defining the Input Headers:
Debt No.

This value works as an identification number for when we define which type of debt finance we will be using.

Debt Structure Types

I/O stands for interest-only, **CA** for constant amortisation, **CP** for constant payment, **CP-B** for constant payment with a balloon, **R-Up** for rolled-up interest, **SWp** for cash sweep.

Interest Rate Type

Fixed or floating. A fixed interest rate is one that doesn't change over the period of the loan, whereas the floating rate will go up and down depending on a benchmark, for example, SONIA (UK), EURIBOR (Euro Zone), and SOFR (US). We will start with a fixed rate for all debt types and then see how to use the floating rate to find the cost of debt.

We would like to create a drop-down menu here, so we go to **Data > Data Validation**

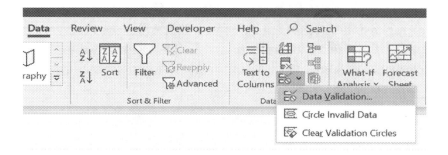

Figure 6.2 Data Validation.

Then, in the **Allow:** drop-down menu**,** click **List**.

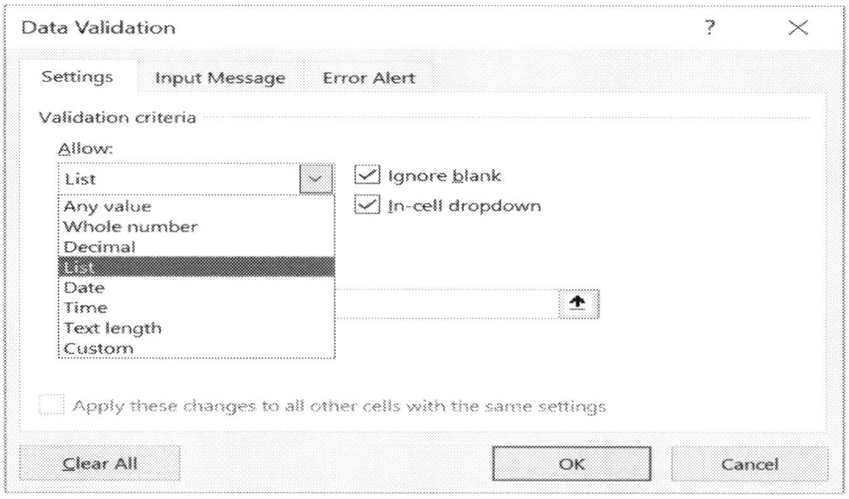

Figure 6.3 Data Validation 2.

Then, in the **Source** field, type **Fixed, Floating**.

Figure 6.4 Data Validation 3.

Then, **OK**.
If you click on this cell, you will see a drop-down menu:

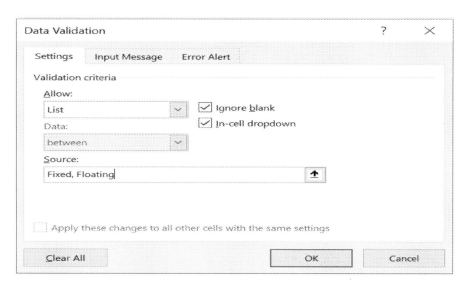

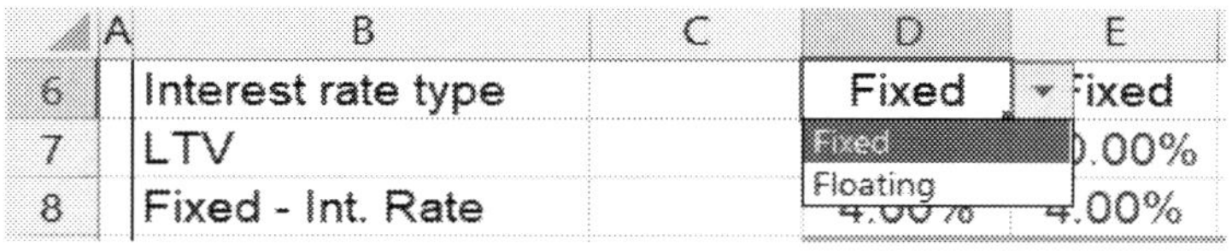

Figure 6.5 Data Validation 4.

Loan to Value (LTV)

Loan to value (or LTV) is the ratio of debt/value of the property. At the start of the borrowing, this is the amount of debt divided by the purchase price of the property. For example, if the property is bought for £10m and the debt finance is £5m, then the LTV is £5/£10 is 0.5 or 50%. As time goes, property values may go up or down depending on the market and the outstanding debt amount may go down depending on debt type and amortisation schedule, so the LTV will change over time.

Fixed – Interest Rate

Self-explanatory: this is the nominal interest rate on a fixed-rate basis.

Floating - Margin

This is the margin above the benchmark rate that will be applied to the debt. When we assume a fixed rate, this cell should be greyed out, so that we know that this is not relevant.

We can do so by using condition formatting:

Home > Conditional Formatting > New Rule

Figure 6.6 ....

Use a formula to determine which cells to format:
Then, enter the formula in the field of **Format values where this formula is true**

   = Interest rate type cell = "Fixed"

Notice here that the cell value has no dollar signs around it ($). Excel will automatically add the dollar signs in the reference cell for the formula; make sure you delete those so we can copy the conditional formatting to the right.

Debt Structures 109

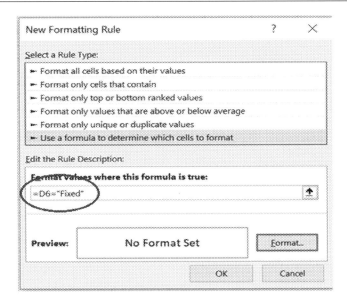

Figure 6.7 Conditional Formatting 2.

Click **Format**.
On **Font** tab, change **Colour.**

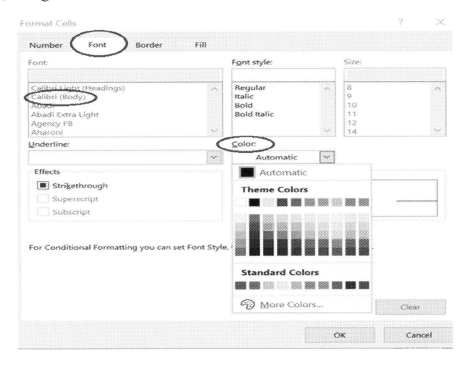

Figure 6.8 Conditional Formatting 3.

Then, on the **Fill** tab, change to the same colour of Font (so it 'disappears'):

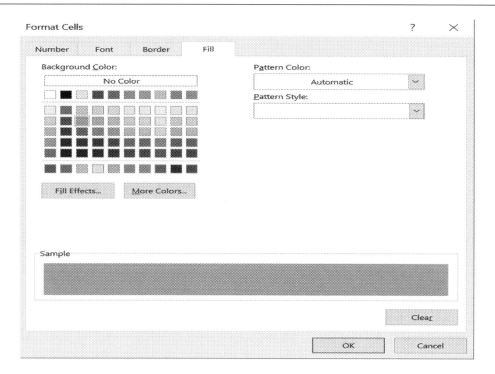

Figure 6.9 Conditional Formatting 4.

Amortisation

Amortisation refers to situations where there is an agreement for interim repayment of the principal amount. For example, the agreement in the interest-only loans is that there will be no amortisation throughout the life of the loan, i.e., the principal amount will only be repaid at the expiry date.

Amortisation Period

This input is only relevant for the debt type constant payment with a balloon. It refers to the timeframe of amortisation which is longer than the maturity. In this case, the debt is not fully amortised, and a final lump sum is due at the end of the debt period.

Maturity

The maturity period is when the loan expires. Typically, the loan expiry will coincide with the sale of the asset. If the maturity is shorter than the holding period, investors will normally refinance the loan. If the maturity is longer than the holding period, investors will repay the debt early.

Fees

Lenders on real estate commonly include various charges and fees in addition to the interest rate as a condition of making a loan. These fees increase the effective cost of borrowing and should be carefully addressed.

Fee amounts vary depending on the economic cycle. For instance, if there is more money supply in the economy, then lenders will compete to supply loans and fees may be reduced and vice versa.

Below, we will describe some of the most common fees and we will model only the origination fees as a starting point. We also add a **Rolled Fee?** switch as Yes/No, so we can model if the fee is paid through equity (not rolled) or if it is incorporated in the loan amount (Yes).

Origination fees (or front-end fees): charged when the loan is made or closed; these are intended to cover expenses incurred by the lender for processing and underwriting loan applications, preparing loan documentation, amortisation schedules, obtaining credit reports, and any other expenses that the lender believes should be recovered by the borrower. The origination fee also includes the broker's fee.

Prepayment fees: it is effectively a penalty fee on the existing loan that must be paid in case the borrower decides to pay the loan in full at an earlier date than the maturity date.

Redemption fees: these fees are due at the time of exiting the loan when the debt has been repaid in full. These are sometimes referred to as exit fees.

Agency fees: this is an annual fee, calculated as a lump sum or per bank per annum, payable by the borrower to compensate the agent for the mechanical and operational work performed by that bank under the loan agreement.

Commitment fees: an annual percentage fee payable to a bank on the undrawn portion of a committed loan facility. Typically paid quarterly in arrears.

Extension fees: a fee charged when an existing committed facility is extended beyond the original maturity date.

Facility fee: annual percentage fee, payable by the borrower pro-rata to banks providing a credit facility to that borrower – it is calculated on the full amount of the facility, whether or not the facility is utilised.

Utilisation fee: fee paid to the lender to increase its return on drawn assets. The payment is generally linked to the average utilisation of the facility exceeding a specified percentage or amount during a defined period of time.

There are many other fees that may be applied to a loan, however, in this book, we will only model the origination fee which is paid at the start of the loan term.

Debt Covenants

To minimise the risks of not receiving their money back, lenders impose many requirements for making the loan. These are called covenants. Some of them may be operational, such as maintaining and insuring the property, but here, we are interested in the financial requirements, mainly:

- Interest coverage ratio (ICR)
- Debt service coverage ratio (DSCR)
- Loan to value (LTV)
- Debt yield

Interest Coverage Ratio (ICR)

The **I**nterest **C**overage **R**atio is an income-based risk measure that will look at the net operating income from the project and compare with the amount of interest that is due in the period.

$$ICR = NOI/Interest\ Due$$

In essence, ICR measures how many times the project could pay its interest payment with its available cash. Some agreements will look at instantaneous ICR ratios, i.e., only at the interest due and net operating income cash flow at the period in question. Some will look four quarters back and create an average, i.e. last year's ICR, and some will look forward four quarters, i.e. next year's ICR. Some will look at longer periods than that. As such the underwriting process for a loan may be bespoke for each lending.

Credit ratings company, such as Moody's, Standard & Poor, and Fitch, will have their own criteria to classify debt instruments. But as bank loans are not listed instruments, then lenders' credit analysis may vary from bank to bank, fund to fund.

To add the 'x' next to the ratio, just press **CTRL + 1 > Number > Custom**
In **Type** field, type: #,##0.00x.

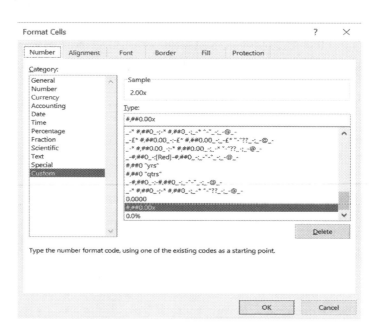

Figure 6.10 Conditional Formatting 5.

Debt Service Coverage Ratio (DSCR)

The debt service coverage ratio is another income-based risk measure that will look at the net operating income from the project and compare with the amount of interest, fee, and amortisation that are due in the period.

As with ICR, the DSCR measures how many times over (or under) the project could pay its total debt obligation with its available cash.

$$DSCR = NOI/(Interest + fee + amortisation)$$

Loan to Value (LTV)

At the starting point of debt underwriting, it is uncommon to see LTV being breached; otherwise, the loan would have not been granted at all. However, as time goes on and the market moves, economic slowdown and uncertainties can make values go down. In this case, incorporating LTV to the analysis will help the lender and the borrower decide on how much volatility in price the property can face before the LTV covenant is breached.

Note: The difference between the previous LTV and this one is that the previous LTV is the ratio of debt to the value of the property, whereas LTV under covenants is the maximum ratio of debt to the value at which the lender can borrow.

Debt Yield

The debt yield ratio is defined as the net operating income (NOI) or net rent divided by the outstanding loan amount. Note that the debt yield considers how large a loan is compared to the property's rent. The idea here is that should the loan default, what would be the lender's year 1 return if the property were to be taken over by the lender.

Debt Yield = NOI/Outstanding Loan

Input Table

After you finish adding some inputs, the first column with your first debt type should look like this

Debt No.	1	2	3	4	5	6
DEBT DETAILS						
Type	I/O	CA	CP	CP-B	R-Up	SWp
Interest rate type	Fixed					
LTV	60.00%					
Fixed - Int. Rate	4.00%					
Floating - Margin						
Amortisation						
Amort. Period						
Maturity	10 yrs					
Fees						
Origination fee	1.00%					
Rolled fee?	Yes					
Covenants						
ICR	2.00x					
DSCR	1.50x					
LTV	70.0%					
Debt Yield	5.00%					

Figure 6.11 Debt Input Table.

You can then highlight the first column and copy to the right to the number 6 Debt Type:

Debt No.	1	2	3	4	5	6
DEBT DETAILS						
Type	I/O	CA	CP	CP-B	R-Up	SWp
Interest rate type	Fixed					
LTV	60.00%					
Fixed - Int. Rate	4.00%					
Floating - Margin						
Amortisation						
Amort. Period						
Maturity	10 yrs					
Fees						
Origination fee	1.00%					
Rolled fee?	Yes					
Covenants						
ICR	2.00x					
DSCR	1.50x					
LTV	70.0%					
Debt Yield	5.00%					

Figure 6.12 Debt Input Table 2.

Press **CTRL + R**.

	A	B	C	D	E	F	G	H	I
2		DEBT DETAILS							
3									
4		Debt No.		1	2	3	4	5	6
5		Type		I/O	CA	CP	CP-B	R-Up	SWp
6		Interest rate type		Fixed	Fixed	Fixed	Fixed	Fixed	Fixed
7		LTV		60.00%	60.00%	60.00%	60.00%	60.00%	60.00%
8		Fixed - Int. Rate		4.00%	4.00%	4.00%	4.00%	4.00%	4.00%
9		Floating - Margin							
10		Amortisation							
11		Amort. Period							
12		Maturity		10 yrs	10 yrs	10 yrs	10 yrs	10 yrs	10 yrs
13									
14		Fees							
15		Origination fee		1.00%	1.00%	1.00%	1.00%	1.00%	1.00%
16		Rolled fee?		Yes	Yes	Yes	Yes	Yes	Yes
17									
18		Covenants							
19		ICR		2.00x	2.00x	2.00x	2.00x	2.00x	2.00x
20		DSCR		1.50x	1.50x	1.50x	1.50x	1.50x	1.50x
21		LTV		70.0%	70.0%	70.0%	70.0%	70.0%	70.0%
22		Debt Yield		5.00%	5.00%	5.00%	5.00%	5.00%	5.00%

Figure 6.13 Debt Input Table 3.

We will leave the same input data for all debt types but will discuss them as we model each structure individually.

Naming Cells

Now that we understand the inputs and have added some values in those, we can start to think about how best to model the debt schedules. One thing I would like to do in this model is to name cells. Naming cells can be particularly helpful when using the same cell in a string of formulas; for example, instead of the formula referring to a cell A5, you will refer to a cell, say, int_rate. The good thing about naming cells is that you can often tell what the cell refers to if you name it in a meaningful way; you can probably tell that int_rate refers to an interest rate.

When naming cells, you should not name a cell with a name that exists already in Excel; for example, you shouldn't name A5, IR20, as IR20 is a cell name already. Cell names cannot have spaces in between so we use _ (underscore) instead of spaces.

Best practice is also not to name a cell a function that exists in Excel, for example, IRR, NPV as it can get quite confusing if you are going to use these functions. Best practice also dictates that you should be mindful of names used: don't use swear words, please!

A point to make on naming cells is that cell-naming has its controversy in financial modelling: some modellers think it is a great idea, some not so much. You can pick your side – just make sure you are consistent throughout, i.e., if you decide to name cells, try to name all cells in a consistent way, for instance, all debt input data with similar names.

To name cells, all you need to do is click on the cell you want to name and type the name on the left hand cell field:

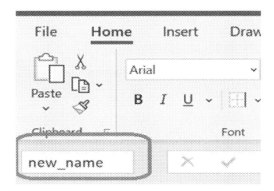

Figure 6.14 Naming Cells.

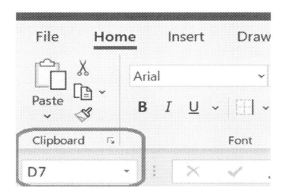

Figure 6.15 Naming Cells 2.

We will name the relevant cells for each debt type as we model them.

Cell Names

The most relevant cells that we will need to name are: **LTV, Fixed-Int. Rate, Maturity, Origination Fee, and Rolled fee?**

We will use the following system to name the cells:

LTV = ltv_No
Fixed-Int. Rate = fixed_IR_No
Maturity = mat_No
Origination Fee = fee_No
Rolled fee? = rolled_fee_No

Note that **No** means number, so for the first debt type, it would be 1, the second, 2, and so on.

Amortisation and **Amort. Period** are only relevant for one type of debt each, constant amortisation and constant payment with a balloon, respectively. In that case, we will not create a system as such.

We will also name on the **Asset** sheet, the Purchase Price cell as gross_price since this is the price inclusive of purchaser's costs.

Debt Structures Modelling

We will now start to model the main debt types and we will first consider that they are all structured as senior debt.

Layout

We will use the layout below for all debt structures. Please note that the asset-level **net operating income** has already been calculated in the **Asset** sheet and it is therefore linked to the **Debt** sheet.

	M	N	O
2	**Net Cash Flow**		
3			
4	Period		0
5	Date		31-Mar-21
6			
7	NOI		13,363
8	NCF		-654,763
9	*Est. Value*		890,833
10	*Fwd Rates*		
11			
12	I/O		1
13	**Senior Debt - Interest Only**		
14	Debt B/F		
15	Debt Drawdown		
16	Interest Due		
17	Interest Paid		
18	Fees Due		
19	Fees Paid		
20	Amortisation		
21	Final Capital Repayment		
22	Debt C/F		
23	Debt Cash Flow		
24			
25	Geared Cash Flow		

Figure 6.16 Debt Structures Model.

Labels and Formulas

DEBT BROUGHT FORWARD

Debt B/F - The opening balance of the debt in the beginning of the period. It is the same as the debt carried forward from the previous period.

> = Debt C/F of previous period

DEBT DRAWDOWN

The amount that is received from the lender. It is the LTV × Purchase Price. Here, I have just mentioned 'Purchase Price', but the amount borrowed can be either net or gross of fees. We will assume that the borrowing amount is based on the gross price, i.e., with purchaser's costs included in this value. Note that if the basis for the LTV is the net asset value (excluding purchaser's costs), then the reference should change to the net price instead.

We need to add two conditionals to the debt drawdown formula, in Excel world, IF statements: One for the initial period (we only drawdown the amount at the start of the debt term, and one for the fees related to the debt facility).

> =(Period=0) × (LTV × Gross Price)/IF (fees rolled into debt amount = "yes", (1 + fee), 1)

INTEREST DUE

The interest accrued from the debt.

> = Debt B/F × Interest Rate × YEARFRAC (start date, end date)

The **YEARFRAC** function gives you the fraction of the years represented by the number of days between the start date and end date.

INTEREST PAID

Sometimes, interest is paid in full, sometimes it is paid as the maximum of available net cash flow (NCF), and sometimes, it is not paid until the maturity date, in case of rolled-up interest debt types.

= Depends!

FEES DUE

> = Loan Amount × Fee Rate

FEES PAID

Fees can be paid from equity (not rolled up into the loan) or embedded in the loan amount (rolled up into the loan).

= Depends!

AMORTISATION

This is the amortised amount in each period. Amortisation will reduce the outstanding debt amount and helps in reducing the risk of default.

= Depends!

FINAL CAPITAL REPAYMENT

This is the amount that will be paid at maturity, i.e., the outstanding balance in the last period of the debt.

> = - IF (final period, Debt B/F, 0)

DEBT CARRIED FORWARD (C/F)

This is the sum of everything above which will be the opening balance in the next period.

DEBT CASH FLOW

This is the sum of debt drawdown, interest paid, fees paid, amortisation, and final capital repayment.

Interest Only (I/O)

As the name suggests, interest-only loans mean that you pay the interest, but throughout the life of the loan, you will not be amortising (repaying) any of the capital; as such, the loan balance remains the same and the whole capital amount is due at the end of the loan term.

For the borrower, this represents less risk in the shorter term since there is certainty on payments – i.e., interest payments are constant if we consider fixed interest rates. On the other hand, if the property value declines dramatically, the borrower may end up in a situation where the proceeds of the sale will not be able to cover the loan amount.

For the lender, the underwriting process focuses not only on the borrower's creditworthiness but also on the end value of the property. The assumption here is that at the end of the loan life, the property will hold its value and the borrower will be able to refinance the debt or pay back the loan. However, refinancing risk is high in turbulent markets and the property may be sold to repay the debt. In this case, the lender will face delay in the repayment due to the illiquidity of real estate assets and lower values.

The interest-only debt cash flow looks like this:

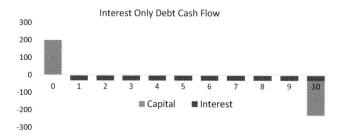

Figure 6.17 Interest Only 1.

Debt Structures 119

EXERCISE 1

Assume a senior debt structure. Based on a I/O structure as per the Debt Type table below, calculate the actual cost of debt (EAR), minimum ICR, DSCR, LTV, debt yield as well as geared IRR and NPV for the project.

Input Data

	B	C	D	E	F	G	H	I
2	**DEBT DETAILS**							
4	Debt No.		1	2	3	4	5	6
5	Type		I/O	CA	CP	CP-B	R-Up	SWp
6	Interest rate type		Fixed	Fixed	Fixed	Fixed	Fixed	Fixed
7	LTV		60.00%	60.00%	60.00%	60.00%	60.00%	60.00%
8	Fixed - Int. Rate		4.00%	4.00%	4.00%	4.00%	4.00%	4.00%
9	Floating - Margin							
10	Amortisation							
11	Amort. Period							
12	Maturity		10 yrs	10 yrs	10 yrs	10 yrs	10 yrs	10 yrs
14	**Fees**							
15	Origination fee		1.00%	1.00%	1.00%	1.00%	1.00%	1.00%
16	Rolled fee?		Yes	Yes	Yes	Yes	Yes	Yes
18	**Covenants**							
19	ICR		2.00x	2.00x	2.00x	2.00x	2.00x	2.00x
20	DSCR		1.50x	1.50x	1.50x	1.50x	1.50x	1.50x
21	LTV		70.0%	70.0%	70.0%	70.0%	70.0%	70.0%
22	Debt Yield		5.00%	5.00%	5.00%	5.00%	5.00%	5.00%

Figure 6.18 Interest Only 2.

Asset Cash Flow

To start with, we need to model the asset cash flow in terms of NOI and NCF as follows:

	M	N	O
2	**Net Cash Flow**		
4	=Asset!AA4		=Asset!AC4
5	=Asset!AA5		=Asset!AC5
7	NOI		=Asset!AC13
8	NCF		=Asset!AC22
9	Est. Value		=SUM(O7:R7)/LOOKUP(YEAR(O5),Asset!$P$10:$P$14,Asset!$S$10:$S$14)

	M	N	O	P	Q	BA	BB	BC
2	**Net Cash Flow**							
4	Period		0	1	2	38	39	40
5	Date		31-Dec-21	31-Mar-22	30-Jun-22	30-Jun-31	30-Sep-31	31-Dec-31
7	NOI		13,363	13,363	13,363	16,279	13,159	-
8	NCF		-654,763	13,363	13,363	16,279	-20,591	800,152
9	Est. Value		890,833	855,200	855,200	420,535	187,980	0

Figure 6.19 Asset Cash Flow.

Then, following the previous explanation, we will need to start creating the formulas for the interest-only debt schedule.

Formulas

Debt B/F = Debt C/F of previous period (= N22)

Debt Drawdown = (Period=0) × (LTV × Gross Price) = (O4 = 0) × ltv_1 × gross_price

Interest Due = Debt B/F × Interest Rate × YEARFRAC (start date, end date) = O14 × fixed_IR_1 × YEARFRAC (N5, O5, 0)

Interest Paid = - (negative) of Interest Due = -O16

Feeds Due = Loan Amount × Fee Rate = fee_1 × O15

Fees Paid = - (Is fee rolled in the loan amount = "No") × Fees Due – (Period = Loan Maturity) × (Fee Due – SUM (Paid Fees)
= - (rolled_fee_1 = "no") × O18 – (O4 = mat_1 × 4) × (O$18 – SUM ($N19:N19))

Note that here we are adding up an empty cell with itself, but the reason why we are doing this is because as we drag the formula do the right, N19 will become O19, P19 and the SUM range will expand accordingly. The idea is to capture the fact that if there was already a fee payment as equity in Period 0, then there will not be a final payment of fee in the maturity period. If you do not do this, the formula will become a circular reference and will not work, so you need to start the SUM with the previous cell and fix the column so that the SUM expands the array as you move along.

Amortisation = 0 (zero). Remember, it is an interest-only facility, so the loan is not amortising and should therefore be zero.

Final Capital Repayment = - IF (final period, Debt B/F + accruals - repayments, 0)
= - IF(O4 = mat_1 × 4) × SUM (O14:O20)

Debt C/F = The SUM of everything above (you can press AUTOSUM, or CTRL + =)
= SUM (O14:O21)

	M	N	O
12	I/O	1	
13	**Senior Debt - Interest Only**		
14	Debt B/F		=N22
15	Debt Drawdown		= (O$4=0)* ltv_1*gross_price
16	Interest Due		=O14*fixed_IR_1*YEARFRAC(N$5,O$5,0)
17	Interest Paid		=-O16
18	Fees Due		=fee_1*O15
19	Fees Paid		=-(rolled_fee_1="no")*O18 - (O$4=mat_1*4)*($O18-SUM($N19:N19))
20	Amortisation		0
21	Final Capital Repayment		=-(O$4=mat_1*4)*SUM(O14:O20)
22	Debt C/F		=SUM(O14:O21)
23	Debt Cash Flow		=O15+O17+O19+O20+O21
24			
25	Geared Cash Flow		=O$8+O23

Figure 6.20 Interest-Only Debt Schedule Formulas.

Check the dollar signs for fixing rows, columns or everything. Highlight cells to the right, and press **CTRL + R**

Covenants

ICR

We will use the forward-looking ICR:

> = **SUM** (NOI over the next four periods)/**SUM** (Interest Due over the next 4 periods)

However, we will need to adjust the formula to cope with structural errors, such as calculating the last four quarters, when the NOI will add zero futures values, since they have not been calculated as the cash flow ends. Therefore, we will add the following **IF** statement to get rid of the errors:

> = **IF** (**AND** (period > 0, period < mat_No × 4 - 3), ICR, "")

With the above IF function, we will make the first and last four periods' ICRs 'disappear'.

DSCR

As per above, the DSCR will be calculated on a forward-looking basis:

> = **SUM** (NOI over the next four periods)/**SUM** (Interest Due over the next 4 periods) – Amortisation Due over the next 4 periods)

We will again add the IF statement to get rid of the errors.

> = **IF** (**AND** (period > 0, period < mat_No × 4 - 3), DSCR, "")

LTV

For us to calculate LTV, we will need to first estimate the values. Estimating values is a very complex art, but mathematically, we can use an initial yield forecast and the NOI as basis for values.

In the Real Estate Investment & Financial Modelling course, our forecast stopped at rental growth rates. Now, we can add an initial yield forecast to it. In the *Asset* sheet, find:

Property Forecast		
Year	Rental Growth	Initial Yield
Year	% p.a.	% p.q.	
2021	-2.00%	-0.50%	6.00%
2022	0.00%	0.00%	6.25%
2023	-1.00%	-0.25%	6.50%
2024	1.50%	0.37%	6.75%
2025	2.00%	0.50%	7.00%

Source: Example Research
Date: Nov-21

Figure 6.21 Property Forecast.

Estimated Values

We will estimate values as NOI/Yield. In Excel:

> = **SUM** (NOI over the next four periods/**LOOKUP** (**YEAR** (cash flow date), Property Forecast Years, Property Forecast Initial Yield)

We will put the estimated values as a line below the NOI so that we can refer to these values in all other debt structures. Once we calculated the estimated property values, the LTV is:.

LTV = Outstanding Debt Amount/Estimated Value

To be consistent, we will use the **IF** statement to get rid of calculation errors.

Debt Yield

> Debt Yield = **SUM** (NOI over the next four periods/Outstanding Debt Amount)

	M	N	O
27	ICR		=IF(AND(O$4>0,O$4<mat_1*4-3), SUM(O$7:R$7)/SUM(O16:R16),"")
28	DSCR		=IF(AND(O$4>0, O$4<mat_1*4-3), SUM(O$7:R$7)/(SUM(O16:R16)-SUM(O20:R20)),"")
29	LTV		=IF(AND(O$4>0, O$4<mat_1*4-3), O22/O$9,"")
30	Debt Yield		=IF(AND(O$4>0, O$4<mat_1*4-3), SUM(O$7:R$7)/O22,"")

	L	M	N	O	P	Q	R	AV	AW	AX	AY	AZ
25												
27	ICR				3.30	3.30	3.30	3.87	3.94	4.02	3.83	
28	DSCR				3.30	3.30	3.30	3.87	3.94	4.02	3.83	

Figure 6.22 – Covenants Calculation

Figure 6.22 Covenants Calculation.

Have Debt Covenants Been Breached?

To answer this question, we will need to look at the calculated covenants and see if any of them has been breached.

Now that we have the debt schedule calculated and all the covenants, we can look at the covenants. If any of the covenants have been breached, it would be a case of the debt structure not being viable, i.e., the lender wouldn't be granting the loan.

To check whether covenants have been breached, we need to look at the MINIMUM value of ICR, DSCR, debt yield, and MAXIMUM value of LTV. If any of these values are lower or greater than they should be, the debt is in a vulnerable position.

	A	B	C	D
25		**Covenants Output**		1
26		Min ICR		=MIN(O27:BC27)
27		Min DSCR		=MIN(O28:BC28)
28		Max LTV		=MAX(O29:BC29)
29		Min Debt Yield		=MIN(O30:BC30)

Figure 6.23 Covenants Formulas.

	A	B	C	D	E	F	G	H	I
17									
18		Covenants							
19		ICR		2.00x	2.00x	2.00x	2.00x	2.00x	2.00x
20		DSCR		1.50x	1.50x	1.50x	1.50x	1.50x	1.50x
21		LTV		70.0%	70.0%	70.0%	70.0%	70.0%	70.0%
22		Debt Yield		5.00%	5.00%	5.00%	5.00%	5.00%	5.00%
23									
24									
25		Covenants Output		1	2	3	4	5	6
26		Min ICR		3.29x					
27		Min DSCR		3.29x					
28		Max LTV		51%					
29		Min Debt Yield		13%					

Figure 6.24 Interest Only 5.

Looking at the table of covenants and covenants output, we can see that none of the covenants were breached, so it would be okay to grant the loan to the borrower under those conditions.

Has Leverage Improved Financial Returns?

To answer this question, we will look at the IRR and NPV of the project, as these will be the ratios that will indicate if debt adds value to the project. Of course, it may well be that the borrower doesn't have enough capital to cover all the acquisition costs; in this case, the only way forward would be to borrow money, but adding debt increases the risk of profile of the project, so debt should be added only when the asset is able to perform on 'its own feet'.

The financial ratios we will calculate are the IRR, NPV, and worth. We will also add the EAR (effective annual rate) to the calculations.

From our outputs, we see that IRR > Target Return and NPV > 0; therefore, we can say that the debt has improved the project's financial performance.

Note that the EAR is 4.18% and not 4.0% as the nominal interest rate. Why? The reason why the EAR differs from the nominal interest rate p.a. is because the borrower will need to consider all the fees, pre-payments and any debt reserve account that can impact the amount of interest paid throughout the life of the loan.

We will also calculate the EAR for different types of debt structures so that we can compare the effective cost of debt across them.

IRR	Geared IRR of the project
	= XIRR (Geared NCF, Cash Flow Dates)
NPV	Net Present Value of the Geared Project
	= XNPV (discount rate, Geared NCF, Cash Flow Dates)
Worth	Goal Seek for NPV = 0 (zero), as calculated above
EAR	Effective Annual Rate
	= XIRR (Debt Cash Flow, Cash Flow Dates)

Figure 6.25 Returns and Cost of Debt Calculation.

	B	C	D
32	**Structures Output**		
33	*Target Return*		0.15
34	Debt No.		1
35	IRR		=XIRR(O25:BC25,O5:BC5)
36	NPV		=XNPV(D33,O25:BC25,O5:BC5)
37	Worth		=D36+gross_price
38	EAR		=XIRR(O23:BC23,O5:BC5)

	B	C	D	E	F	G	H	I
32	**Structures Output**							
33	*Target Return*		15.00%	15.00%	15.00%	15.00%	15.00%	15.00%
34	Debt No.		1	2	3	4	5	6
35	IRR		18.35%					
36	NPV		45,053					
37	Worth		713,178					
38	EAR		4.18%					

Figure 6.26 EAR Effective 1.

Outputs Calculations

	A	B	C	D
25		**Covenants Output**		1
26		Min ICR		3.29x
27		Min DSCR		3.29x
28		Max LTV		51%
29		Min Debt Yield		13%
30				
31				
32		**Structures Output**		
33		*Target Return*		15.00%
34		Debt No.		1
35		IRR		18.35%
36		NPV		45,053
37		Worth		713,178
38		EAR		4.18%

	B	C	D
25	**Covenants Output**		1
26	Min ICR		=MIN(O27:BC27)
27	Min DSCR		=MIN(O28:BC28)
28	Max LTV		=MAX(O29:BC29)
29	Min Debt Yield		=MIN(O30:BC30)
30			
31			
32	**Structures Output**		
33	*Target Return*		0.15
34	Debt No.		1
35	IRR		=XIRR(O25:BC25,O5:BC5)
36	NPV		=XNPV(D33,O25:BC25,O5:BC5)
37	Worth		=D36+gross_price
38	EAR		=XIRR(O23:BC23,O5:BC5)

Figure 6.27 EAR Effective 2.

Interest-Only Structure – Complete Cash Flow Formulas

	M	N	O
2	**Net Cash Flow**		
3			
4	=Asset!AA4		=Asset!AC4
5	=Asset!AA5		=Asset!AC5
6			
7	NOI		=Asset!AC13
8	NCF		=Asset!AC22
9	Est. Value		=SUM(O7:R7)/LOOKUP(YEAR(O5),Asset!$P$10:$P$14,Asset!$S$10:$S$14)
11			
12	I/O	1	
13	**Senior Debt - Interest Only**		
14	Debt B/F		=N22
15	Debt Drawdown		= (O$4=0)* ltv_1*gross_price
16	Interest Due		=O14*fixed_IR_1*YEARFRAC(N$5,O$5,0)
17	Interest Paid		=-O16
18	Fees Due		=fee_1*O15
19	Fees Paid		=-(rolled_fee_1="no")*O18 - (O$4=mat_1*4)*($O18-SUM($N19:N19))
20	Amortisation		0
21	Final Capital Repayment		=-(O$4=mat_1*4)*SUM(O14:O20)
22	Debt C/F		=SUM(O14:O21)
23	Debt Cash Flow		=O15+O17+O19+O20+O21
24			
25	Geared Cash Flow		=O$8+O23
26			
27	ICR		=IF(AND(O$4>0,O$4<mat_1*4-3), SUM(O$7:R$7)/SUM(O16:R16),"")
28	DSCR		=IF(AND(O$4>0,O$4<mat_1*4-3), SUM(O$7:R$7)/(SUM(O16:R16)-SUM(O20:R20)),"")
29	LTV		=IF(AND(O$4>0, O$4<mat_1*4-3), O22/O$9,"")
30	Debt Yield		=IF(AND(O$4>0, O$4<mat_1*4-3), SUM(O$7:R$7)/O22,"")

Figure 6.28 Interest-Only Debt Structure – Complete Cash Flow Formulas.

The results of the calculation will be

	M	N	O	P	Q	R	AV	AW	AX	AY	AZ	BA	BB	BC
2	**Net Cash Flow**													
3														
4	Period		0	1	2	3	33	34	35	36	37	38	39	40
5	Date		31-Dec-21	31-Mar-22	30-Jun-22	30-Sep-22	31-Mar-30	30-Jun-30	30-Sep-30	31-Dec-30	31-Mar-31	30-Jun-31	30-Sep-31	31-Dec-31
6														
7	NOI		13,363	13,363	13,363	13,363	15,038	15,038	16,279	16,279	16,279	16,279	13,159	-
8	NCF		-854,763	13,363	13,363	13,363	15,038	15,038	16,279	16,279	16,279	16,279	-20,991	800,152
9	Est. Value		890,833	855,200	855,200	855,200	894,779	912,500	930,221	885,645	653,090	420,535	187,980	0
11														
12	I/O	1												
13	**Senior Debt - Interest Only**													
14	Debt B/F		-	404,884	404,884	404,884	404,884	404,884	404,884	404,884	404,884	404,884	404,884	404,884
15	Debt Drawdown		400,875	-	-	-	-	-	-	-	-	-	-	-
16	Interest Due		-	4,049	4,049	4,049	4,049	4,049	4,049	4,049	4,049	4,049	4,049	4,049
17	Interest Paid		-	-4,049	-4,049	-4,049	-4,049	-4,049	-4,049	-4,049	-4,049	-4,049	-4,049	-4,049
18	Fees Due		4,009	-	-	-	-	-	-	-	-	-	-	-
19	Fees Paid		-	-	-	-	-	-	-	-	-	-	-	-4,009
20	Amortisation		-	-	-	-	-	-	-	-	-	-	-	-
21	Final Capital Repayment		-	-	-	-	-	-	-	-	-	-	-	-400,875
22	Debt C/F		404,884	404,884	404,884	404,884	404,884	404,884	404,884	404,884	404,884	404,884	404,884	-
23	Debt Cash Flow		400,875	-4,049	-4,049	-4,049	-4,049	-4,049	-4,049	-4,049	-4,049	-4,049	-4,049	-408,933
24														
25	Geared Cash Flow		-253,888	9,314	9,314	9,314	10,990	10,990	12,230	12,230	12,230	12,230	-24,640	391,220
26														
27	ICR			3.30	3.30	3.30	3.87	3.94	4.02	3.83				
28	DSCR			3.30	3.30	3.30	3.87	3.94	4.02	3.83				
29	LTV			47.34%	47.34%	47.34%	45.25%	44.37%	43.53%	45.72%				
30	Debt Yield			13.20%	13.20%	13.20%	15.47%	15.78%	16.08%	15.31%				

Figure 6.29 Interest-Only Complete Cash Flow Values.

Amortising Loans

Until now, we explained how to do the data input and modelling of the debt schedule of interest-only loans. Now, the focus will shift towards a different approach – amortised real estate debt arrangements. Unlike interest-only scenarios where the principal remains the same until maturity, amortised loans involve regular payments that will reduce the principal over time.

Using the same debt scheduling structure and data input table used before, we will show you how to model a a) fully amortising, constant amortisation loan, b) constant amortisation with a constant amount with a ballon repayment, c) constant amortisation with a constant percentage of the outstanding amount with a ballon repayment, d) fully amortising, constant payment (annuity), and finally, e) constant payment with a ballon payment.

Note that even though I am including fully amortising loans in this chapter, this is not a type of debt that is seen in commercial real estate, mainly because loans tend to be for a maximum of ten years term, and by fully amortising the loan over this period, it would be a negative cash flow and will be consistently generated over the investment period which is not optimal. However, the reason for adding it here is because individual mortgages may have this feature of full amortisation, but in this case, the term of the loan would be much longer, between 20 and 30 years depending on the age of the borrower.

Fully Amortising – Constant Amortisation

In this type of debt arrangement, the borrower (the investor buying the asset) will drawdown the loan amount as they acquire the property and repay a constant amount or percentage of the outstanding debt balance as amortisation. The interest amount will go down from period to period, and at the end of the term, the borrower should have paid off all of the loan.

The amortisation amount per quarter can be calculated as:

Amortisation amount = LTV × gross price × (amortisation rate/4)

This will mean that the total amount will be amortised over the life of the loan, leaving a zero balance at the end of the loan term.

So, if your loan is for £1,000,000 and the tenor is 5 years, the amortising amount would be:

Amort = 1,000,000/(5 × 4) = 200,000

The fully amortising constant amortisation mortgage cash flow looks like this:

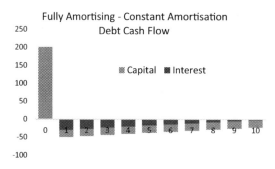

Figure 6.30 Constant Amortisation.

Partially Amortising – Constant Amount

Say you agree with the lender that the amortisation will be constant at 2% p.a. of the initial loan for the same 5 years. Then, the amortising amount would be:

Amort = 1,000,000 × 2%/4 = 5,000

The cash flow of the partially amortising, constant amount, and constant amortisation mortgage looks like this:

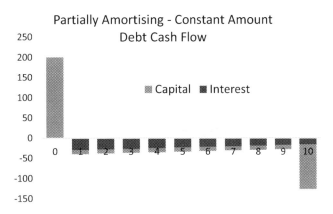

Figure 6.31 Constant Amortisation 2.

Partially Amortising – Constant Percentage (CA)

Lenders can also agree to charge a percentage based on the outstanding loan balance; say 2% p.a. or 0.50% per quarter.

The amortising amount would be:

Period 1 = 1,000,000 × 0.50% = 5,000
Period 2 = Outstanding balance = 1,000,000 − 5,000 (Period 1 amortisation)
= 995,000 × 0.50% = 4,975

The cash flow of the partially amortising, constant percentage of the outstanding loan balance looks like this:

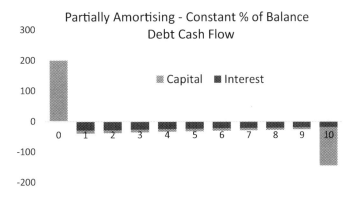

Figure 6.32 Constant Amortisation 3.

Depending on which type of amortisation you use, the risk may be very similar to an interest-only mortgage, i.e., high re-financing risk or risk of default at the end of the loan if the property value declines considerably.

The chart below will give you an idea of the outstanding debt balance using the three methods of constant amortisation explained and the idea is that the greater the balance at the end of the debt term, the higher the re-financing or default risk.

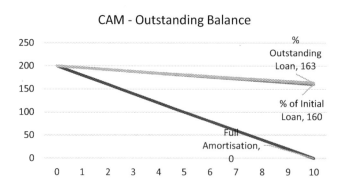

Figure 6.33 Constant Percentage.

As such, if the amortisation schedule used is one that fully amortises the loan, then the re-financing risk will disappear. However, the high initial mortgage payments will not normally match the property income available to service the debt and interim default is therefore high.

In the model for this book, we will not use a partially amortising loan with a percentage of the outstanding balance as amortisation. We will however learn how to create a fully amortising loan using the constant payment mortgage in the next section.

EXERCISE 2

Assume a senior debt structure. Based on a constant amortisation structure as per the Debt Type table below, calculate the actual cost of debt (EAR), minimum ICR, DSCR, LTV, debt yield as well as geared IRR and NPV for the project. The amortisation percentage is based on a constant amount, i.e., the percentage of the initial borrowing amount.

Constant Amortisation Debt Structure Modelling

Input Data

Now that we are more familiar with the model, we follow the same pattern and add the details for the **constant amortisation debt structure**.

DEBT DETAILS							
Debt No.		1	2	3	4	5	6
Type		I/O	CA	CP	CP-B	R-Up	SWp
Interest rate type		Fixed	Fixed	Fixed	Fixed	Fixed	Fixed
LTV		60.00%	60.00%	60.00%	60.00%	60.00%	60.00%
Fixed - Int. Rate		4.00%	4.00%	4.00%	4.00%	4.00%	4.00%
Floating - Margin							
Amortisation			2.00%				
Amort. Period							
Maturity		10 yrs	10 yrs	10 yrs	10 yrs	10 yrs	10 yrs
Fees							
Origination fee		1.00%	1.00%	1.00%	1.00%	1.00%	1.00%
Rolled fee?		Yes	Yes	Yes	Yes	Yes	Yes
Covenants							
ICR		2.00x	2.00x	2.00x	2.00x	2.00x	2.00x
DSCR		1.50x	1.50x	1.50x	1.50x	1.50x	1.50x
LTV		70.0%	70.0%	70.0%	70.0%	70.0%	70.0%
Debt Yield		5.00%	5.00%	5.00%	5.00%	5.00%	5.00%

Figure 6.34 Input Data.

Renaming Cells

	A	B	C	D	E	F	G	H	I
2		**DEBT DETAILS**							
3									
4		Debt No.		1	2	3	4	5	6
5		Type		I/O	CA	CP	CP-B	R-Up	SWp
6		Interest rate type		Fixed	Fixed	Fixed	Fixed	Fixed	Fixed
7		LTV		60.00%	ltv_2	60.00%	60.00%	60.00%	60.00%
8		Fixed - Int. Rate		4.00%	fixed_IR_2	4.00%	4.00%	4.00%	4.00%
9		Floating - Margin							
10		Amortisation			amort_2				
11		Amort. Period							
12		Maturity		10 yrs	mat_2	10 yrs	10 yrs	10 yrs	10 yrs
13									
14		Fees							
15		Origination fee		1.00%	fee_2	1.00%	1.00%	1.00%	1.00%
16		Rolled fee?		Yes	rolled_fee_2	Yes	Yes	Yes	Yes
17									

Figure 6.35 Input Data 2.

Fixing Cells from I/O Debt Structure

What we would like to do is to replicate the I/O structure to all other structures that will come after. So, go back to the I/O schedule and fix the rows that are related to the NOI/NCF or estimated value. The affected cells are highlighted in yellow below.

Add a $ to all references to rows BEFORE the start of the Debt B/F cell:

	L	M	N	O
11				
12		I/O	1	
13		**Senior Debt - Interest Only**		
14		Debt B/F		=N22
15		Debt Drawdown		= (O$4=0)* ltv_1*gross_price
16		Interest Due		=O14*fixed_IR_1*YEARFRAC(N$5,O$5,0)
17		Interest Paid		=-O16
18		Fees Due		=fee_1*O15
19		Fees Paid		=-(rolled_fee_1="no")*O18 - (O$4=mat_1*4)*($O18-SUM($N19:N19))
20		Amortisation		0
21		Final Capital Repayment		=-(O$4=mat_1*4)*SUM(O14:O20)
22		Debt C/F		=SUM(O14:O21)
23		Debt Cash Flow		=O15+O17+O19+O20+O21
24				
25		Geared Cash Flow		=O$8+O23
26				
27		ICR		=IF(AND(O$4>0,O$4<mat_1*4-3), SUM(O$7:R$7)/SUM(O16:R16),"")
28		DSCR		=IF(AND(O$4>0, O$4<mat_1*4-3), SUM(O$7:R$7)/(SUM(O16:R16)-SUM(O20:R20)),"")
29		LTV		=IF(AND(O$4>0, O$4<mat_1*4-3), O22/O$9,"")
30		Debt Yield		=IF(AND(O$4>0, O$4<mat_1*4-3), SUM(O$7:R$7)/O22,"")

Figure 6.36 Fixing Cells.

Copying and Pasting to the CA Debt Schedule:

Highlight the first column of the I/O Debt Schedule, copy, and paste onto the CA Debt Schedule.

Change the references of cell names, for example, from ltv_1 to ltv_2 and so on.

130 Real Estate Financial Modelling in Excel

	M	N	O
33	CA	2	
34	**Senior Debt - Constant Amortisation**		
35	Debt B/F		=N43
36	Debt Drawdown		= (O$4=0)* ltv_1*gross_price
37	Interest Due		=O35*fixed_IR_1*YEARFRAC(N$5,O$5,0)
38	Interest Paid		=-O37
39	Fees Due		=fee_1*O36
40	Fees Paid		=-(rolled_fee_1="no")*O39 - (O$4=mat_1*4)*($O39-SUM($N40:N40))
41	Amortisation		0
42	Final Capital Repayment		=-(O$4=mat_1*4)*SUM(O35:O41)
43	Debt C/F		=SUM(O35:O42)
44	Debt Cash Flow		=O36+O38+O40+O41+O42
45			
46	Geared Cash Flow		=O$8+O44

Figure 6.37 Debt Schedule.

Changing the Cell References (Note that the only difference will be changing the references from 1 to 2, e.g. ltv_1 to ltv_2 etc):

	L	M	N	O
32				
33		CA	2	
34		**Senior Debt - Constant Amortisation**		
35		Debt B/F		=N43
36		Debt Drawdown		= (O$4=0)* ltv_2*gross_price
37		Interest Due		=O35*fixed_IR_2*YEARFRAC(N$5,O$5,0)
38		Interest Paid		=-O37
39		Fees Due		=fee_2*O36
40		Fees Paid		=-(rolled_fee_2="no")*O39 - (O$4=mat_2*4)*($O39-SUM($N40:N40))
41		Amortisation		
42		Final Capital Repayment		=-(O$4=mat_2*4)*SUM(O35:O41)
43		Debt C/F		=SUM(O35:O42)
44		Debt Cash Flow		=O36+O38+O40+O41+O42
45				
46		Geared Cash Flow		=O$8+O44
47				
48		ICR		=IF(AND(O$4>0,O$4<mat_2*4-3), SUM(O$7:R$7)/SUM(O37:R37),"")
49		DSCR		=IF(AND(O$4>0, O$4<mat_2*4-3), SUM(O$7:R$7)/(SUM(O37:R37)-SUM(O41:R41)),"")
50		LTV		=IF(AND(O$4>0, O$4<mat_2*4-3), O43/O$9,"")
51		Debt Yield		=IF(AND(O$4>0, O$4<mat_2*4-3), SUM(O$7:R$7)/O43,"")

Figure 6.38 Cell Reference.

Constant Amortisation (%) Formula

The only difference between an I/O structure and a constant amortisation one is that there will be some capital repayment (amortisation) during the life of the loan.

In this case, it will be:

Amortisation = - amortisation rate/period × LTV × price

We will add two conditionals (TRUE or FALSE), so that the amortisation 'appears' only at period 1 onwards (and not period 0) and does not 'appear' in the last period, because then the amortisation will be part of the final capital repayment.

Debt Structures

	M	N	O
33	CA	2	
34	**Senior Debt - Constant Amortisation**		
35	Debt B/F		=N43
36	Debt Drawdown		= (O$4=0)* ltv_2*gross_price
37	Interest Due		=O35*fixed_IR_2*YEARFRAC(N$5,O$5,0)
38	Interest Paid		=-O37
39	Fees Due		=fee_2*O36
40	Fees Paid		=-(rolled_fee_2="no")*O39 - (O$4=mat_2*4)*($O39-SUM($N40:N40))
41	Amortisation		=- (O$4>0) * (O$4<mat_2*4) * $O$36*amort_2/4
42	Final Capital Repayment		=-(O$4=mat_2*4)*SUM(O35:O41)
43	Debt C/F		=SUM(O35:O42)
44	Debt Cash Flow		=O36+O38+O40+O41+O42
45			
46	Geared Cash Flow		=O$8+O44

Figure 6.39 Amortisation Formula.

Constant Amortisation – Constant Amount Debt Cash Flow

	M	N	O	P	Q	R	S	BB	BC
33	CA	2							
34	**Senior Debt - Constant Amortisation**								
35	Debt B/F		-	404,884	402,879	400,875	398,871	328,718	326,713
36	Debt Drawdown		400,875	-	-	-	-	-	-
37	Interest Due		-	4,049	4,029	4,009	3,989	3,287	3,267
38	Interest Paid		-	-4,049	-4,029	-4,009	-3,989	-3,287	-3,267
39	Fees Due		4,009	-	-	-	-	-	-
40	Fees Paid		-	-	-	-	-	-	-4,009
41	Amortisation		-	-2,004	-2,004	-2,004	-2,004	-2,004	-
42	Final Capital Repayment		-	-	-	-	-	-	-322,704
43	Debt C/F		404,884	402,879	400,875	398,871	396,866	326,713	-
44	Debt Cash Flow		400,875	-6,053	-6,033	-6,013	-5,993	-5,292	-329,980
45									
46	Geared Cash Flow		-253,888	7,309	7,329	7,349	7,369	-25,883	470,172

Figure 6.40 Constant Amortisation.

Constant Payment – Full Amortisation (CP)

Constant payment, or CP, is the lending method whereby the borrower fully amortises the debt, thereby paying a constant amount each period.

Because the principal is fully amortised over the life of the loan, CP is not optimal for property borrowers as the payment level normally exceeds property income. However, residential buyers will mainly be faced with this repayment method given that individual's credit ratings are significantly lower than corporate's ratings in normal market conditions.

The benefit of CP is that there is no re-financing or final capital payment risk; therefore, from a lender's perspective, this is the optimal repayment method.

The constant payment mortgage cash flow looks like this:

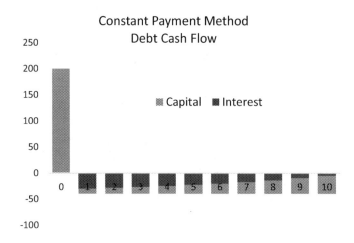

Figure 6.41 Constant Payment.

EXERCISE 3

Assume a senior debt structure. Based on a constant payment structure as per the Debt Type table below, calculate the actual cost of debt (EAR), minimum ICR, DSCR, LTV, debt yield as well as geared IRR and NPV for the project. This is a fully amortising constant payment loan.

The technique is the same as with IO and CA, so all that needs doing is adding the inputs to the debt detail:

A	B	C	D	E	F	G	H	I
2	DEBT DETAILS							
3								
4	Debt No.		1	2	3	4	5	6
5	Type		I/O	CA	CP	CP-B	R-Up	SWp
6	Interest rate type		Fixed	Fixed	Fixed	Fixed	Fixed	Fixed
7	LTV		60.00%	60.00%	60.00%	60.00%	60.00%	60.00%
8	Fixed - Int. Rate		4.00%	4.00%	4.00%	4.00%	4.00%	4.00%
9	Floating - Margin							
10	Amortisation			2.00%				
11	Amort. Period							
12	Maturity		10 yrs	10 yrs	10 yrs	10 yrs	10 yrs	10 yrs
13								
14	Fees							
15	Origination fee		1.00%	1.00%	1.00%	1.00%	1.00%	1.00%
16	Rolled fee?		Yes	Yes	Yes	Yes	Yes	Yes
17								
18	Covenants							
19	ICR		2.00x	2.00x	2.00x	2.00x	2.00x	2.00x
20	DSCR		1.50x	1.50x	1.50x	1.50x	1.50x	1.50x
21	LTV		70.0%	70.0%	70.0%	70.0%	70.0%	70.0%
22	Debt Yield		5.00%	5.00%	5.00%	5.00%	5.00%	5.00%

Figure 6.42 Constant Payment Debt Structure.

For ease, we have already named the cells for you according to our known naming structure.

You just need to copy and paste the formulas from the IO or CA structure to the CP structure, change the cell references from 2 to 3, and change the amortisation formula again.

	M	N	O
54	CP	3	
55	**Senior Debt - Constant Payment**		
56	Debt B/F		=N64
57	Debt Drawdown		= (O$4=0)* ltv_3*gross_price
58	Interest Due		=O56*fixed_IR_3*YEARFRAC(N$5,O$5,0)
59	Interest Paid		=-O58
60	Fees Due		=fee_3*O57
61	Fees Paid		=-(rolled_fee_3="no")*O60 - (O$4=mat_3*4)*($O60-SUM($N61:N61))
62	Amortisation		
63	Final Capital Repayment		=-(O$4=mat_3*4)*SUM(O56:O62)
64	Debt C/F		=SUM(O56:O63)
65	Debt Cash Flow		=O57+O59+O61+O62+O63
66			
67	Geared Cash Flow		=O$8+O65
68			
69	ICR		=IF(AND(O$4>0,O$4<mat_3*4-3), SUM(O$7:R$7)/SUM(O58:R58),"")
70	DSCR		=IF(AND(O$4>0, O$4<mat_3*4-3), SUM(O$7:R$7)/(SUM(O58:R58)-SUM(O62:R62)),"")
71	LTV		=IF(AND(O$4>0, O$4<mat_3*4-3), O64/O$9,"")
72	Debt Yield		=IF(AND(O$4>0, O$4<mat_3*4-3), SUM(O$7:R$7)/O64,"")

Figure 6.43 Constant Payment Debt Structure 2.

Annuity Calculation – PMT

The constant payment debt schedule is one that the borrower pays an annuity (constant amount). As such, we will use the PMT function in Excel that will calculate this annuity for each period. This payment is formed of both interest and amortisation and is fixed in value at each period. The syntax in Excel is:

= PMT (interest rate, debt term, borrowed amount)

Calculating the Amortisation Amount

As we calculated the total payment amount per quarter using the **PMT** function, we only need to subtract the value of the interest paid so we can find the amortisation amount.

We also need to remember to add the conditions that there will only be amortisation when the period is greater than zero, but the amortisation will be paid up until the maturity date, as there will not be a final balloon payment.

> Amortisation = (period > 0) × (period < = maturity) × PMT (interest rate/4, maturity, LTV × gross price) – Interest Paid)

	M	N	O	P	Q	R	AX	AY	AZ	BA	BB	BC
54	CP	3										
55	**Senior Debt - Constant Payment**											
56	Debt B/F		-	404,884	396,602	388,237	71,464	59,848	48,115	36,265	24,297	12,209
57	Debt Drawdown		400,875									
58	Interest Due		-	4,049	3,966	3,882	715	598	481	363	243	122
59	Interest Paid		-	-4,049	-3,966	-3,882	-715	-598	-481	-363	-243	-122
60	Fees Due		4,009	-	-	-	-	-	-	-	-	-
61	Fees Paid		-	-	-	-	-	-	-	-	-	-
62	Amortisation		-	-8,282	-8,365	-8,449	-11,616	-11,733	-11,850	-11,968	-12,088	-12,209
63	Final Capital Repayment											-0
64	Debt C/F		404,884	396,602	388,237	379,788	59,848	48,115	36,265	24,297	12,209	-
65	Debt Cash Flow		400,875	-12,331	-12,331	-12,331	-12,331	-12,331	-12,331	-12,331	-12,331	-12,331
66												
67	Geared Cash Flow		-253,888	1,032	1,032	1,032	3,948	3,948	3,948	3,948	-32,922	787,821
68												
69	ICR			3.41	3.48	3.56	30.19	38.79				
70	DSCR			1.08	1.08	1.08	1.32	1.26				
71	LTV			46.38%	45.40%	44.41%	6.43%	5.43%				
72	Debt Yield			13.48%	13.77%	14.07%	108.80%	128.85%				

Figure 6.44 Amortisation Amount.

134 Real Estate Financial Modelling in Excel

	M	N	O
54	CP	3	
55	**Senior Debt - Constant Payment**		
56	Debt B/F		=N64
57	Debt Drawdown		= (O$4=0)* ltv_3*gross_price
58	Interest Due		=O56*fixed_IR_3*YEARFRAC(N$5,O$5,0)
59	Interest Paid		=-O58
60	Fees Due		=fee_3*O57
61	Fees Paid		=-(rolled_fee_3="no")*O60 - (O$4=mat_3*4)*($O60-SUM($N61:N61))
62	Amortisation		=(O$4>0)*(O$4<=mat_3*4)*(PMT(fixed_IR_3/4,mat_3*4,$O57+$O60)-O59)
63	Final Capital Repayment		=-(O$4=mat_3*4)*SUM(O56:O62)
64	Debt C/F		=SUM(O56:O63)
65	Debt Cash Flow		=O57+O59+O61+O62+O63
66			
67	Geared Cash Flow		=O$8+O65

Figure 6.45 Amortisation Amount 2.

Constant Payment – Balloon (CP-B)

Constant payment debt structure with a balloon payment is the repayment method where the borrower partially amortises the debt but retaining the feature of a fixed amount repayment schedule.

CP with a balloon payment (i.e., the amount amortised does not cover the full loan) can be adopted, with a 30 years' repayment schedule, though the loan term is less than that.

The constant payment debt with balloon repayment cash flow can be demonstrated graphically as the chart below:

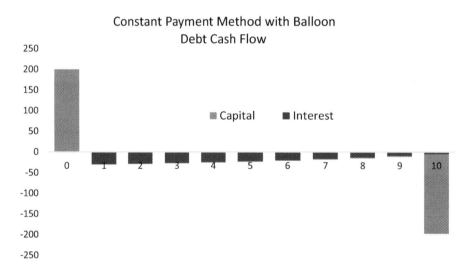

Figure 6.46 Balloon Repayment.

EXERCISE 4

Assume a senior debt structure. Based on a constant payment with a balloon structure as per the Debt Type table below, calculate the actual cost of debt (EAR), minimum ICR, DSCR, LTV, debt yield as well as geared IRR and NPV for the project.

Constant Payment with Balloon Debt Structure Modelling

Input data:

Note that the amortisation period is now 30 years, meaning that although the borrower will pay back the loan at year 10, the interim payments will not be sufficient to fully amortise the loan.

For ease, we have again named the cells for you according to our known naming structure.

You just need to copy and paste the formulas from the structure above, change the cell references, and change the amortisation formula.

	A	B	C	D	E	F	G	H	I
2		DEBT DETAILS							
3									
4		Debt No.		1	2	3	4	5	6
5		Type		I/O	CA	CP	CP-B	R-Up	SWp
6		Interest rate type		Fixed	Fixed	Fixed	Fixed	Fixed	Fixed
7		LTV		60.00%	60.00%	60.00%	60.00%	60.00%	60.00%
8		Fixed - Int. Rate		4.00%	4.00%	4.00%	4.00%	4.00%	4.00%
9		Floating - Margin							
10		Amortisation			2.00%				
11		Amort. Period					30 yrs		
12		Maturity		10 yrs	10 yrs	10 yrs	10 yrs	10 yrs	10 yrs
13									
14		Fees							
15		Origination fee		1.00%	1.00%	1.00%	1.00%	1.00%	1.00%
16		Rolled fee?		Yes	Yes	Yes	Yes	Yes	Yes
17									
18		Covenants							
19		ICR		2.00x	2.00x	2.00x	2.00x	2.00x	2.00x
20		DSCR		1.50x	1.50x	1.50x	1.50x	1.50x	1.50x
21		LTV		70.0%	70.0%	70.0%	70.0%	70.0%	70.0%
22		Debt Yield		5.00%	5.00%	5.00%	5.00%	5.00%	5.00%

Figure 6.47 Balloon Repayment Input Data.

Copying, Pasting and Changing References:

	M	N	O
75	CP-B	4	
76	**Senior Debt - Constant Payment with Balloon**		
77	Debt B/F		=N85
78	Debt Drawdown		= (O$4=0)* ltv_4*gross_price
79	Interest Due		=O77*fixed_IR_4*YEARFRAC(N$5,O$5,0)
80	Interest Paid		=-O79
81	Fees Due		=fee_4*O78
82	Fees Paid		=-(rolled_fee_4="no")*O81 - (O$4=mat_4*4)*($O81-SUM($N82:N82))
83	Amortisation		
84	Final Capital Repayment		=-(O$4=mat_4*4)*SUM(O77:O83)
85	Debt C/F		=SUM(O77:O84)
86	Debt Cash Flow		=O78+O80+O82+O83+O84
87			
88	Geared Cash Flow		=O$8+O86
89			
90	ICR		=IF(AND(O$4>0,O$4<mat_4*4-3), SUM(O$7:R$7)/SUM(O79:R79),"")
91	DSCR		=IF(AND(O$4>0, O$4<mat_4*4-3), SUM(O$7:R$7)/(SUM(O79:R79)-SUM(O83:R83)),"")
92	LTV		=IF(AND(O$4>0, O$4<mat_4*4-3), O85/O$9,"")
93	Debt Yield		=IF(AND(O$4>0, O$4<mat_4*4-3), SUM(O$7:R$7)/O85,"")

Figure 6.48 Balloon Repayment Copying and Pasting.

Constant Payment with Balloon Amortisation Formula

Now, instead of an amortisation period equal to the maturity period in the PMT formula, the amortisation period will have a specific data input. The specific amortising period will be longer than the actual debt term, hence the balloon payment at the end of the period.

> Amortisation = (period > 0) × (period < = maturity) × **PMT** (interest rate/4, amortisation period, LTV × gross price) − Interest Paid)

Figure 6.49 The Amortisation Formula 1.

Figure 6.50 The Amortisation Formula 2.

Rolled-Up Interest

Rolled-up, or capitalised, interest structures are typically used in development projects or when individuals want to release equity from their property.

In the first case, i.e., development, the interest is accrued (i.e., increase the loan amount) and the amortisation is zero. The final outstanding loan balance and accumulated interest are then either refinanced or paid off from the sales proceeds.

In the second case, i.e., the individual, often a retiree will receive part of the value of its property (the LTV) and will not pay interest or capital on the loan. The expectation is that once the individual dies, the property will be sold, and the proceeds will cover both the loan and the accumulated interest.

The rolled-up interest debt cash flow can be graphically illustrated as below:

Figure 6.51 R-Up Cash Flow.

It is therefore apparent that the risk is mainly related to the property value at the end of debt term.

EXERCISE 5

Assume a senior debt structure. Based on a rolled-up interest structure as per the Debt Type table below, calculate the actual cost of debt (EAR), minimum ICR, DSCR, LTV, debt yield as well as geared IRR and NPV for the project.

Rolled-Up Interest Debt Structure Modelling

Input data:

	B	C	D	E	F	G	H	I
2	DEBT DETAILS							
3								
4	Debt No.		1	2	3	4	5	6
5	Type		I/O	CA	CP	CP-B	R-Up	SWp
6	Interest rate type		Fixed	Fixed	Fixed	Fixed	Fixed	Fixed
7	LTV		60.00%	60.00%	60.00%	60.00%	60.00%	60.00%
8	Fixed - Int. Rate		4.00%	4.00%	4.00%	4.00%	4.00%	4.00%
9	Floating - Margin							
10	Amortisation			2.00%				
11	Amort. Period					30 yrs		
12	Maturity		10 yrs	10 yrs	10 yrs	10 yrs	10 yrs	10 yrs
13								
14	Fees							
15	Origination fee		1.00%	1.00%	1.00%	1.00%	1.00%	1.00%
16	Rolled fee?		Yes	Yes	Yes	Yes	Yes	Yes
17								
18	Covenants							
19	ICR		2.00x	2.00x	2.00x	2.00x	2.00x	2.00x
20	DSCR		1.50x	1.50x	1.50x	1.50x	1.50x	1.50x
21	LTV		70.0%	70.0%	70.0%	70.0%	70.0%	70.0%
22	Debt Yield		5.00%	5.00%	5.00%	5.00%	5.00%	5.00%

Figure 6.52 R-Up Data Input.

Copying, Pasting and Changing References:
You will also need to delete the 'Interest Paid' formula and hardcode amortisation as zero (0), since there will be no interim capital repayments.

138 Real Estate Financial Modelling in Excel

	M	N	O
96	R-Up	5	
97	**Senior Debt - Rolled Up Interest**		
98	Debt B/F		=N106
99	Debt Drawdown		= (O$4=0)* ltv_5*gross_price
100	Interest Due		=O98*fixed_IR_5*YEARFRAC(N$5,O$5,0)
101	Interest Paid		
102	Fees Due		=fee_5*O99
103	Fees Paid		=-(rolled_fee_5="no")*O102 - (O$4=mat_5*4)*($O102-SUM($N103:N103))
104	Amortisation		0
105	Final Capital Repayment		=-(O$4=mat_5*4)*SUM(O98:O104)
106	Debt C/F		=SUM(O98:O105)
107	Debt Cash Flow		=O99+O101+O103+O104+O105
108			
109	Geared Cash Flow		=O$8+O107
110			
111	ICR		=IF(AND(O$4>0,O$4<mat_5*4-3), SUM(O$7:R$7)/SUM(O100:R100),"")
112	DSCR		=IF(AND(O$4>0,O$4<mat_5*4-3), SUM(O$7:R$7)/(SUM(O100:R100)-SUM(O104:R104)),"")
113	LTV		=IF(AND(O$4>0,O$4<mat_5*4-3), O106/O$9,"")
114	Debt Yield		=IF(AND(O$4>0,O$4<mat_5*4-3), SUM(O$7:R$7)/O106,"")

Figure 6.53 R-Up Copy/Paste.

Rolled-Up Interest Payment Formula

The borrower will only pay back the debt at maturity, including the accrued interest. Thus:

$$\text{Interest Paid} = - (\text{period} = \text{maturity}) \times \textbf{SUM} \text{ of rolled-up interest}$$

Pay attention to the array and dollar sign ($) in the **SUM** function.

	M	N	O
95			
96	R-Up	5	
97	**Senior Debt - Rolled Up Interest**		
98	Debt B/F		=N106
99	Debt Drawdown		= (O$4=0)* ltv_5*gross_price
100	Interest Due		=O98*fixed_IR_5*YEARFRAC(N$5,O$5,0)
101	Interest Paid		=-(O4=mat_5*4)*SUM($O$100:O100)
102	Fees Due		=fee_5*O99
103	Fees Paid		=-(rolled_fee_5="no")*O102 - (O$4=mat_5*4)*($O102-SUM($N103:N103))
104	Amortisation		0
105	Final Capital Repayment		=-(O$4=mat_5*4)*SUM(O98:O104)
106	Debt C/F		=SUM(O98:O105)
107	Debt Cash Flow		=O99+O101+O103+O104+O105
108			
109	Geared Cash Flow		=O$8+O107

Figure 6.54 R-Up Interest Payment Formula 1.

	M	N	O	P	Q	R	AX	AY	AZ	BA	BB	BC
95												
96	R-Up	5										
97	**Senior Debt - Rolled Up Interest**											
98	Debt B/F		-	404,884	408,933	413,022	567,381	573,559	579,295	585,088	590,939	596,848
99	Debt Drawdown		400,875	-	-	-	-	-	-	-	-	-
100	Interest Due		-	4,049	4,089	4,130	5,679	5,736	5,793	5,851	5,909	5,968
101	Interest Paid		-	-	-	-	-	-	-	-	-	-197,933
102	Fees Due		4,009	-	-	-	-	-	-	-	-	-
103	Fees Paid		-	-	-	-	-	-	-	-	-	-4,009
104	Amortisation		-	-	-	-	-	-	-	-	-	-
105	Final Capital Repayment		-	-	-	-	-	-	-	-	-	-400,875
106	Debt C/F		404,884	408,933	413,022	417,152	573,559	579,295	585,088	590,939	596,848	-
107	Debt Cash Flow		400,875	-	-	-	-	-	-	-	-	-602,817
108												
109	Geared Cash Flow		-253,588	13,363	13,363	13,363	16,279	16,279	16,279	16,279	-20,591	197,335
110												
111	ICR			3.25	3.22	3.19	2.82	2.66				
112	DSCR			3.25	3.22	3.19	2.82	2.66				
113	LTV			47.82%	48.30%	48.78%	61.66%	65.41%				
114	Debt Yield			13.07%	12.94%	12.81%	11.35%	10.70%				

Figure 6.55 R-Up Interest Payment Formula 2.

Cash Sweep

Cash sweep is a debt structure where the borrower has to use all the excess cash (i.e., positive net cash flow from the project) to pay back the debt.

On the downside, the borrower will not have any discretion to use their excess cash. For example, if the borrower sees other investment opportunities that may yield a high rate of return, they are not allowed to divert the cash for this purpose.

On the upside, cash sweeps will tolerate negative cash flows more than other senior debt types since the borrower can roll up interest and capital to the future. Also, if the borrower can repay the debt at a quicker rate, the amount of interest to be paid will be lower in total.

Generally, this type of debt repayment is *not* common, because it would mean that the lender would need access to the borrower's accounts to check the excess cash available; however, it is featured in this chapter because in case of default, a cash sweep clause may be activated and the borrower will need to use all or partial excess cash to accelerate the debt repayment.

The graph below shows how a cash sweep debt cash flow can look like. The pattern of course depends on the project cash flow and the important feature is the volatility of the cash outflows.

Figure 6.56 SWp Cash Flow.

EXERCISE 6

Assume a senior debt structure. Based on cash sweep structure as per the Debt Type table below, calculate the actual cost of debt (EAR), minimum ICR, DSCR, LTV, debt yield as well as geared IRR and NPV for the project.

Cash Sweep Debt Structure Modelling

Input data:

	A B C	D	E	F	G	H	I
2	DEBT DETAILS						
3							
4	*Debt No.*	*1*	*2*	*3*	*4*	*5*	*6*
5	**Type**	**I/O**	**CA**	**CP**	**CP-B**	**R-Up**	**SWp**
6	Interest rate type	Fixed	Fixed	Fixed	Fixed	Fixed	Fixed
7	LTV	60.00%	60.00%	60.00%	60.00%	60.00%	60.00%
8	Fixed - Int. Rate	4.00%	4.00%	4.00%	4.00%	4.00%	4.00%
9	Floating - Margin						
10	Amortisation		2.00%				
11	Amort. Period				30 yrs		
12	Maturity	10 yrs	10 yrs	10 yrs	10 yrs	10 yrs	10 yrs
13							
14	**Fees**						
15	Origination fee	1.00%	1.00%	1.00%	1.00%	1.00%	1.00%
16	Rolled fee?	Yes	Yes	Yes	Yes	Yes	Yes
17							
18	**Covenants**						
19	ICR	2.00x	2.00x	2.00x	2.00x	2.00x	2.00x
20	DSCR	1.50x	1.50x	1.50x	1.50x	1.50x	1.50x
21	LTV	70.0%	70.0%	70.0%	70.0%	70.0%	70.0%
22	Debt Yield	5.00%	5.00%	5.00%	5.00%	5.00%	5.00%

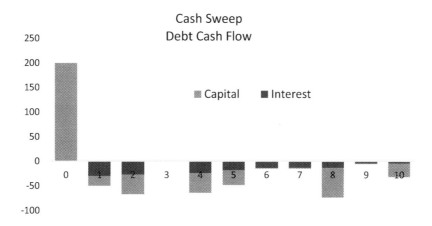

Figure 6.57 SWp Data Input.

Copying, Pasting, and Changing References:

You will need to delete the 'Interest Paid' and 'Amortisation' formulas as these will need to be rewritten for this structure.

You will need to change the second argument of the AND function of the covenants (ICR, DSCR, LTV, and debt yield) as it is very likely that the debt will be paid earlier than maturity. The IF function will then be:

> = IF (AND (period > 0, Balance Brought Forward > 0), Covenant, "")

	M	N	O
117	Swp	6	
118	**Senior Debt - Cash Sweep**		
119	Debt B/F		=N127
120	Debt Drawdown		= (O$4=0)* ltv_6*gross_price
121	Interest Due		=O119*fixed_IR_6*YEARFRAC(N$5,O$5,0)
122	Interest Paid		
123	Fees Due		=fee_6*O120
124	Fees Paid		=-(rolled_fee_6="no")*O123 - (O$4=mat_6*4)*($O123-SUM($N124:N124))
125	Amortisation		
126	Final Capital Repayment		=-(O$4=mat_6*4)*SUM(O119:O125)
127	Debt C/F		=SUM(O119:O126)
128	Debt Cash Flow		=O120+O122+O124+O125+O126
129			
130	Geared Cash Flow		=O$8+O128
131			
132	ICR		=IF(AND(O$4>0,O$119>0), SUM(O$7:R$7)/SUM(O121:R121),"")
133	DSCR		=IF(AND(O$4>0, O$119>0), SUM(O$7:R$7)/(SUM(O121:R121)-SUM(O125:R125)),"")
134	LTV		=IF(AND(O$4>0, O$119>0), O127/O$9,"")
135	Debt Yield		=IF(AND(O$4>0, O$119>0), SUM(O$7:R$7)/O127,"")

Figure 6.58 SWp Copy/Paste.

Cash Sweep Interest Payment Formula (MIN and MAX Functions)

The amount of interest paid in a cash sweep is what the borrower can afford, i.e., it's only paid if there is positive cash flow. So, if the project cash flow is positive, then you pay the **lower** of interest due and available cash.

Lower of in Excel language is MIN. Thus, you can say:

> Interest Paid = MIN (NCF, interest due)

However, we still need to account for 'if the project cash flow is positive'. In this case, we can use MAX for the **greater** of NCF and Zero.

Putting both arguments together:

> Interest Paid = - MAX (MIN (NCF, interest due), 0)

Cash Sweep Amortisation Formula

The amortisation formula will have to reflect the lower of outstanding loan balance minus what have been paid in interest and fees and available cash from the project. Again, this is only possible if there is positive cash flow, i.e., greater of cash flow and zero.

Thus:

> Amortisation = - MAX (MIN (NCF - Interest Paid – Fees Paid, Loan B/F), 0)

	M	N	O
117	Swp	6	
118	**Senior Debt - Cash Sweep**		
119	Debt B/F		=N127
120	Debt Drawdown		=(O$4=0)* ltv_6*gross_price
121	Interest Due		=O119*fixed_IR_6*YEARFRAC(N$5,O$5,0)
122	Interest Paid		=-MAX(MIN(O8,O121),0)
123	Fees Due		=fee_6*O120
124	Fees Paid		=-(rolled_fee_6="no")*O123 - (O$4=mat_6*4)*($O123-SUM($N124:N124))
125	Amortisation		=-MAX(MIN(O8+O122+O124,O119),0)
126	Final Capital Repayment		=-(O$4=mat_6*4)*SUM(O119:O125)
127	Debt C/F		=SUM(O119:O126)
128	Debt Cash Flow		=O120+O122+O124+O125+O126
129			
130	Geared Cash Flow		=O$8+O128

Figure 6.59 SWp Min and Max 1.

	M	N	O	P	Q	R	AX	AY	AZ	BA	BB	BC
117	Swp	6										
118	**Senior Debt - Cash Sweep**											
119	Debt B/F		-	404,884	395,570	386,163	-	-	-	-	-	-
120	Debt Drawdown		400,875	-	-	-	-	-	-	-	-	-
121	Interest Due		-	4,049	3,956	3,862	-	-	-	-	-	-
122	Interest Paid		-	-4,049	-3,956	-3,862	-	-	-	-	-	-
123	Fees Due		4,009	-	-	-	-	-	-	-	-	-
124	Fees Paid		-	-	-	-	-	-	-	-	-	-
125	Amortisation		-	-9,314	-9,407	-9,501	-	-	-	-	-	-
126	Final Capital Repayment		-	-	-	-	-	-	-	-	-	-
127	Debt C/F		404,884	395,570	386,163	376,662	-	-	-	-	-	-
128	Debt Cash Flow		400,875	-13,363	-13,363	-13,363	-	-	-	-	-	-
129												
130	Geared Cash Flow		-253,888	-	-	-	16,279	16,279	16,279	16,279	-20,591	800,152
131												
132	ICR			3.42	3.50	3.59						
133	DSCR			1.00	1.00	1.00						
134	LTV			46.25%	45.15%	44.04%						
135	Debt Yield			13.51%	13.84%	14.19%						

Figure 6.60 SWp Min and Max 2.

Floating Interest Rates

Interest rates charged on the loan do not necessarily need to be fixed as we have assumed so far. The debt can be based on a floating rate – if interest rates go up, the debt will become more expensive and vice versa. Floating rates are based on short-term interest rates which will vary depending on the state of the economy and the effects of government intervention, such as quantitative easing policies.

Short-term interest rates are the rates at which short-term borrowings are affected between financial institutions (for example, SONIA in the UK, EURIBOR in the Euro Zone, and SOFR in the USA). Short-term interest rates are generally averages of daily rates, measured as a percentage but expressed as an annual figure. Short-term interest rates are based on three-month money market rates where available.

Together with SOFR (secured overnight financing rate), SONIA is currently the UK's transaction-based interest rate produced by the Bank of England, and it represents the average cost of borrowing pounds sterling from other banks and financial institutions overnight and used for calculating floating rates, the interest on transactions, and valuation of various assets.

When a debt instrument is structured using a floating rate, the borrower will pay the short-term interest rate plus a margin.

The margin will depend on the borrower's credit rating and vary according to economic cycles. For example, if a company is rated by a credit rating agency to AAA, this margin will be significantly lower than a company rated B.

Projects can also be rated if set up independently in that lenders will have recourse over the project's assets or the borrower's overall assets. In the case of listed bonds, this will be analysed based on its own merits depending on the collateral or guarantees by the issuers.

Commercial mortgage-backed securities (CMBS) also have ratings separated from the sponsor, and in fact, depending on how the CMBS is structured with their different subordination and tranches, the same pool of properties can be credited with different ratings.

Debt structures based on floating rates can have a negative impact for the borrower on their cash flow if interest rates go up; consequently, most borrowers would want to make use of financial instruments such as interest rate swaps and other derivatives to maintain a certain stability in their interest payments when faced with the risk of increased interest rates.

The Yield Curve

The yield curve shows the relationship between yields or interest rates (y or vertical axis) and the time to maturity (x or horizontal axis). These interest rates are rates at which buyers of bonds will be willing to lend the money to the government over a period of time. If you were to lend to someone else, for example, to a real estate project, you will need a premium to take on more risk than that of the government. As such, the yield curve will give you the risk-free rate, because governments would not default on their own debt denominated in their own currency as they have the ability to 'print' more money. In case of government debt issued in a different currency, say US dollars or Euros, then these bonds are not considered the risk-free asset in their jurisdictions.

The yield curve works for both long- and short-term interest rates. If you are borrowing or lending at a fixed rate, then you will be looking at the longer-term rates, but if you are looking at the floating rate, then you will certainly be focussing on the short-term rates, one- to three-month rates.

The longer the time to maturity, the higher the rate of return which explains an upward sloping of the curve in 'normal' economic conditions, i.e., when the market is not envisaging a downturn or recession.

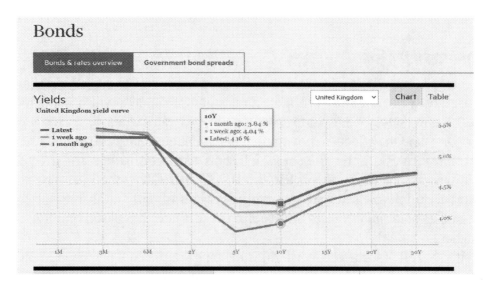

Figure 6.61 Bonds.
Source: FT.com.

Bank Rate

The most important interest rate in the UK is the bank rate, similar to the US Federal Funds Rate and the European Central Bank Deposit Interest Rates. These rates, set by monetary policy, are the interest rates paid to commercial banks that hold money with the Central Banks.

Interest rates charged by commercial or retail banks on money borrowed or saved is influenced by the bank rate. The higher the bank rate, the higher the interest rates charged on individual borrowing and vice versa. This affects consumer spending and eventually has an influence on prices and inflation.

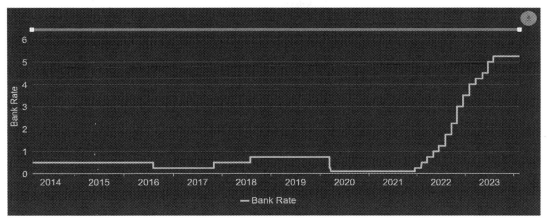

Figure 6.62 Bank Rate.
Source: Bank of England.

Short-Term Interest Rates

The rate at which short-term borrowings are made between banks and other financial institutions is the short-term interest rates: also known as the money market rate or treasury bill rate. Although the bank rates can seem very stable, short-term interest rates do vary as depicted in the below figure. Short-term interest rates are not static, and economic downturns, such as the latest Global Financial Crisis in 2008/2009, Brexit, pandemics, wars, and energy crisis, can have a massive impact on interest rates.

Short-term Interest Rates, Total, % per annum, Nov 1999 to Nov 2023

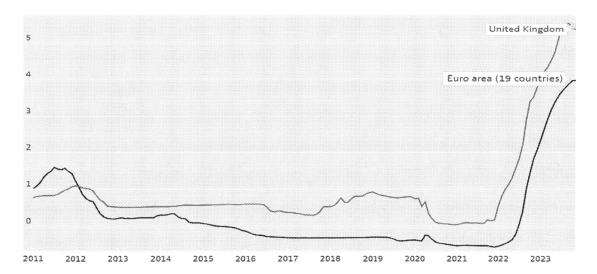

Figure 6.63 Short-Term Interest Rates.
Source: OECD.

Interest Rates Forecast

Short-term interest rate forecast is the estimated value of three-month rates calculated as a percentage. Forecasting short-term interest rates or the Bank of England base rate is not an easy task (no forecast is!). Investment banks, central banks, and other financial institutions will have their own forecast teams and forecast data is calculated based on the overall economic climate in the individual country and worldwide and can be heavily statistical.

The forecast below is for the Bank of England base rate, but as we now know, this has a very important effect on short-term commercial rates.

As you will see in the figure of forecasted bank rate in the UK, the forecast by Statista from 2017, and even though the forecast for 2020 was for the Bank of England rate, to be above 1%, no-one could have predicted a pandemic in 2020 when the bank rate was virtually zero, and in 2024, it being above 5%.

Forecasted Bank Rate in the United Kingdom (UK) from 1st Quarter 2017 to 1st Quarter 2024

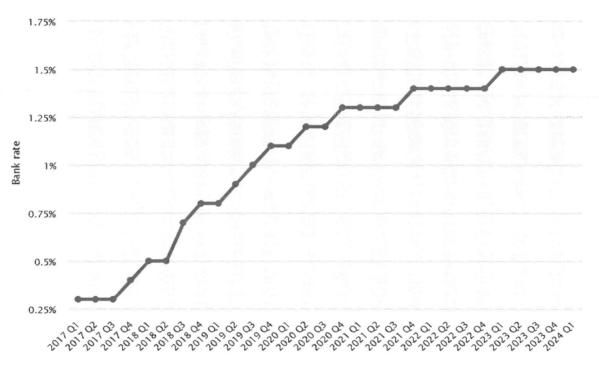

Figure 6.64 Forecasted Bank Rate.
Source: Statista.

The Forward Curve

The forward curve shows the different projections of short-term rates, in the UK, the SONIA, and in the US, the SOFR, which is influenced by the current rates. So, if you can borrow today for 1 year at a rate of 5.0% and for 2 years at a rate of 6.0%, then you can say that in one year's time, the 1-year rate will be 7.01%, because $1.06 \times 1.06 = 1.05 \times (1+ x)$. Solving for x, we have 7.01%.

Consequently, if you take the yield curve which is the government interest rate today (the 'spot' curve), then you can apply the so-called bootstrapping and derive short-term rates based on the current observable yields.

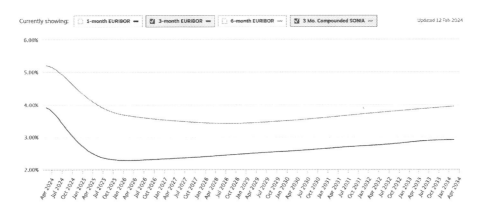

Figure 6.65 Forward Curve.
Source: Chatham Financial.

The forward curve can be used as the 'best' forecast for short-term interest rates we have today but remember that these figures can and will change constantly and considerably, and you should make sure that your models have a mechanism to update the forward curve on a consistent basis.

Covenants and Margins

Now that we know about interest rates, the next input is margins. The margin is also known as premium or default spread, meaning that the higher the risk, the higher the margin and total interest rate paid by the borrower.

Floating Interest Rates Debt Structure Modelling

Exercise 8

Assume a senior debt structure. Based on an interest-only structure as per the Debt Type table below and floating interest rate with a 3.0% margin, calculate the actual cost of debt (EAR), minimum ICR, DSCR, LTV, debt yield as well as geared IRR and NPV for the project.

Calculate the EAR, minimum ICR, DSCR and project IRR and NPV.

Input data:
From Exercise 7, we will change the following inputs:

		1	2	3	4	5	6
DEBT DETAILS							
Debt No.		1	2	3	4	5	6
Type		I/O	CA	CP	CP-B	R-Up	SWp
Interest rate type		Floating	Fixed	Fixed	Fixed	Fixed	Fixed
LTV		60.00%	60.00%	60.00%	60.00%	60.00%	60.00%
Fixed - Int. Rate		4.00%	4.00%	4.00%	4.00%	4.00%	4.00%
Floating - Margin		3.00%					
Amortisation			2.00%				
Amort. Period					30 yrs		
Maturity		10 yrs	10 yrs	10 yrs	10 yrs	10 yrs	10 yrs
Fees							
Origination fee		1.00%	1.00%	1.00%	1.00%	1.00%	1.00%
Rolled fee?		Yes	Yes	Yes	Yes	Yes	Yes
Covenants							
ICR		2.00x	2.00x	2.00x	2.00x	2.00x	2.00x
DSCR		1.50x	1.50x	1.50x	1.50x	1.50x	1.50x
LTV		70.0%	70.0%	70.0%	70.0%	70.0%	70.0%
Debt Yield		5.00%	5.00%	5.00%	5.00%	5.00%	5.00%

Figure 6.66 Floating Interest Rate 1.

Let's call the floating – margin cell **margin_1.** And the interest rate type **int_type_1.**

We can create an intermediate calculation of floating rates using the LOOKUP function somewhere in your cash flow:

= LOOKUP (YEAR (cash flow date), years vector, interest rate vector)

	L	M	N	O
11				
12		I/O	1	=LOOKUP(YEAR(O$5),Asset!$P$9:$P$14,Asset!$R$10:$R$14)
13		Senior Debt - In		
14		Debt B/F		-
15		Debt Drawdown		400,875
16		Interest Due		-
17		Interest Paid		-
18		Fees Due		4,009
19		Fees Paid		-
20		Amortisation		-
21		Final Capital Repaym		-
22		Debt C/F		404,884
23		Debt Cash Flow		400,875
24				
25		Geared Cash Flow		-253,888

Figure 6.67 Floating Interest Rate 2.

	M	N										
12	I/O	1	0.00%	-0.25%	-0.25%	-0.25%	-0.25%	0.37%	0.37%	0.00%	0.00%	0.00%
13	Senior Debt - Interes											
14	Debt B/F		-	404,884	404,884	404,884	404,884	404,884	404,884	404,884	404,884	404,884
15	Debt Drawdown		400,875	-	-	-	-	-	-	-	-	-
16	Interest Due		-	4,049	4,049	4,049	4,049	4,049	4,049	4,049	4,049	4,049
17	Interest Paid		-	-4,049	-4,049	-4,049	-4,049	-4,049	-4,049	-4,049	-4,049	-4,049
18	Fees Due		4,009	-	-	-	-	-	-	-	-	-
19	Fees Paid		-	-	-	-	-	-	-	-	-	-4,009
20	Amortisation		-	-	-	-	-	-	-	-	-	-
21	Final Capital Repayme		-	-	-	-	-	-	-	-	-	-400,875
22	Debt C/F		404,884	404,884	404,884	404,884	404,884	404,884	404,884	404,884	404,884	-
23	Debt Cash Flow		400,875	-4,049	-4,049	-4,049	-4,049	-4,049	-4,049	-4,049	-4,049	-408,933
24												
25	Geared Cash Flow		-253,888	9,314	9,314	9,314	9,314	9,314	9,314	12,230	-24,640	391,220
26												
27	ICR			3.30	3.30	3.30	3.30	3.30	3.30			
28	DSCR			3.30	3.30	3.30	3.30	3.30	3.30			
29	LTV			47.34%	47.34%	47.34%	47.34%	49.26%	49.29%			
30	Debt Yield			13.20%	13.20%	13.20%	13.20%	13.19%	13.19%			

Figure 6.68 Floating Interest Rate 3.

Now, we need to adjust the Interest Due formula, because that will be the only line to have any changes in the structure. Currently, the formula is:

Interest Due = Balance B/F × Fixed Interest Rate × YEARFRAC

However, our interest rate is not fixed anymore, but variable. In this case, we will add a condition (IF) function to say:

= IF (Interest Rate Type = "Fixed", Fixed Interest Rate, Floating Rate + Margin)

Putting together both formulas of fixed and floating interest rates, we have:

Interest Due = Balance B/F × IF (Interest Rate Type = "Fixed", Fixed Interest Rate, Floating Rate + Margin) × YEARFRAC

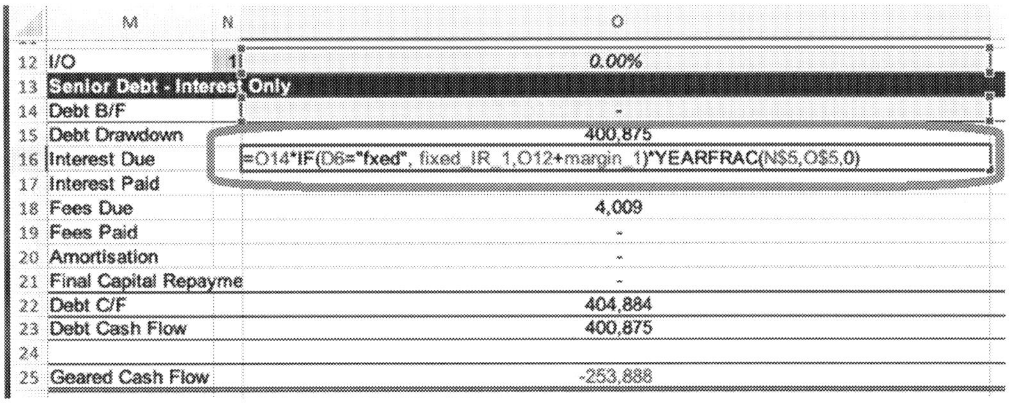

Figure 6.69 Floating Interest Rate 4.

Copy and paste to the right. Don't be tempted to think that Excel can read your mind and copy the cell to the right just because you entered a new formula in your first reference cell.

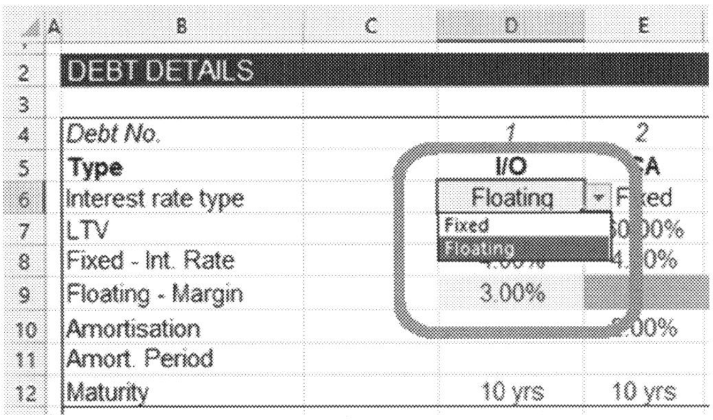

Change the interest type to floating:

Figure 6.70 Floating Interest Rate 5.

When changing the interest type to floating, you get:

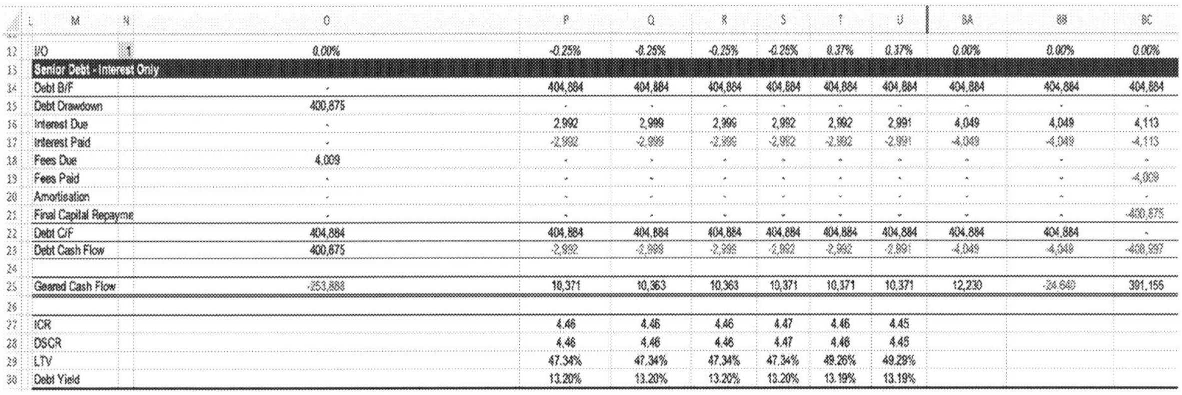

Figure 6.71 Floating Interest Rate 6.

You can do the exact same change to all debt structures, from interest-only to cash sweep; but in the interest of time, we will leave this change for you to do when you go through this model again as you study the topic. It can also be that you would like to compare two debt facilities that have the same structure, for example, interest only, but one is fixed and the other is floating. In this case, you will need to 'copy and paste' the relevant debt schedule as a separate structure, with a new debt number, and you also need to adjust the 'Live' cash flow to include this new structure.

Conclusion

In this chapter, we modelled geared real estate investments, using senior debt finance. I showed you how to create a series of debt schedules and covenants calculations and applied the most common types of debt repayments, namely the interest-only, constant amortisation, constant payment, rolled-up interest, and cash sweep structures. We also looked at fixed and floating interest rates and how to model these features into the cash flow.

This chapter also taught us how to create practical elements such as input tables and dynamic formulas using elegant debt modelling techniques in Excel. By the end of this chapter, you should have a solid understanding of how different debt structures can be modelled, how to manage covenants, and be able to identify any 'red flags' that gearing can bring to the real estate investment.

In the next chapter, I will show you how to create a 'live' debt schedule, how to analyse the outputs of the geared investment, and find the maximum loan amount.

Chapter 7

Comparing Debt Structures and Analysing Results

Chapter Contents

Introduction	149
Covenants and Geared Returns	152
Calculate 'Live' Covenants and Returns	152
Create a Data Table	153
Finding the Maximum LTV	154
Geared vs. Ungeared Returns	156
Linking Chosen to Live Returns	156
The Output Table	158
Sensitivity Analysis	158
Credit Enhancements	159
Debt Service Reserve Account (Cash Reserve)	159
Cash Reserve Modelling	159
Revolving Facility Modelling	162
Upfront Fee	162
Commitment Fee	162
Utilisation Fee	162
Expiry	163
Layout	163
Loan Drawdown	163
Principal Repayment	163
Interest + Fees Paid	163
Conclusion	164

Introduction

Given the options of senior debt discussed in the previous chapter, you may be asking yourself: 'so what?' I know, there are different types of debt, different covenants and we need to pick one that will fit our investment strategy. In this case, my solution is to create a mechanism that will automatically work out the outputs given the different types of debt and lending options. My solution is creating a 'live' debt schedule and geared cash flow that will reflect the 'chosen' structure – interest-only, constant amortisation, etc.

We will therefore replicate the debt schedule and use the **CHOOSE** function to link it with the debt schedules calculated in the previous chapter.

First, you will need to copy the debt schedule structure as we have done before. Please refer to the previous chapter if you are not familiar with the 'copy and paste' feature of debt scheduling.

Then, we will use the **CHOOSE function.**

The **CHOOSE** function returns a value from a list using a given position. For example, **CHOOSE**(2, "red", "blue", "green") returns "blue", since blue is the second value listed after the index number.

In our case, because our debt structure identifications are cardinal numbers (1, 2, 3, and so on), we will choose the corresponding value of each structure.

	L	M	N	O
138		Live		=Asset!D30
139		Senior Debt - Cash Sweep		
140		Debt B/F		=CHOOSE($N$138,O14,O35,O56,O77,O98,O119)
141		Debt Drawdown		
142		Interest Due		
143		Interest Paid		
144		Fees Due		
145		Fees Paid		
146		Amortisation		
147		Final Capital Repayment		
148		Debt C/F		
149		Debt Cash Flow		
150				
151		Geared Cash Flow		
152				
153		ICR		
154		DSCR		
155		LTV		
156		Debt Yield		

Figure 7.1 Choose Function.

We can then highlight all cells below until Debt Yield and press **CTRL + D**.

	M	N	O
138	Live	=Asset!D30	
139	Senior Debt - Cash Sweep		
140	Debt B/F		=CHOOSE($N$138, O14,O35,O56,O77,O98,O119)
141	Debt Drawdown		=CHOOSE($N$138, O15,O36,O57,O78,O99,O120)
142	Interest Due		=CHOOSE($N$138, O16,O37,O58,O79,O100,O121)
143	Interest Paid		=CHOOSE($N$138, O17,O38,O59,O80,O101,O122)
144	Fees Due		=CHOOSE($N$138, O18,O39,O60,O81,O102,O123)
145	Fees Paid		=CHOOSE($N$138, O19,O40,O61,O82,O103,O124)
146	Amortisation		=CHOOSE($N$138, O20,O41,O62,O83,O104,O125)
147	Final Capital Repayment		=CHOOSE($N$138, O21,O42,O63,O84,O105,O126)
148	Debt C/F		=CHOOSE($N$138, O22,O43,O64,O85,O106,O127)
149	Debt Cash Flow		=CHOOSE($N$138, O23,O44,O65,O86,O107,O128)
150			=CHOOSE($N$138, O24,O45,O66,O87,O108,O129)
151	Geared Cash Flow		=CHOOSE($N$138, O25,O46,O67,O88,O109,O130)
152			=CHOOSE($N$138, O26,O47,O68,O89,O110,O131)
153	ICR		=CHOOSE($N$138, O27,O48,O69,O90,O111,O132)
154	DSCR		=CHOOSE($N$138, O28,O49,O70,O91,O112,O133)
155	LTV		=CHOOSE($N$138, O29,O50,O71,O92,O113,O134)
156	Debt Yield		=CHOOSE($N$138, O30,O51,O72,O93,O114,O135)

Figure 7.2 Choose Function 2.

Then, you can highlight all the way to the right and **CTRL + R**.

Figure 7.3 Choose Function 3.

The problem now is that we lost the formatting:

Figure 7.4 Choose Function 4.

So, the best option is to drag the formulas down, and once done, there will appear a small box on the bottom right corner. If you click on it, you can choose to **Fill Without Formatting**.

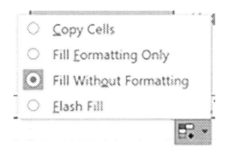

Figure 7.5 Fill without Formatting.

Also, don't forget to delete the rows which should be empty. There you go.

	M	N	O	P	Q	R	AX	AY	AZ	BA	BB	BC
138	Live		6									
139	**Senior Debt - Cash Sweep**											
140	Debt B/F			404,884	395,570	386,163	-	-	-	-	-	-
141	Debt Drawdown		400,875	-	-	-	-	-	-	-	-	-
142	Interest Due		-	4,049	3,956	3,862	-	-	-	-	-	-
143	Interest Paid		-	-4,049	-3,956	-3,862	-	-	-	-	-	-
144	Fees Due		4,009	-	-	-	-	-	-	-	-	-
145	Fees Paid		-	-	-	-	-	-	-	-	-	-
146	Amortisation		-	-9,314	-9,407	-9,501	-	-	-	-	-	-
147	Final Capital Repayment		-	-	-	-	-	-	-	-	-	-
148	Debt C/F		404,884	395,570	386,163	376,662	-	-	-	-	-	-
149	Debt Cash Flow		400,875	-13,363	-13,363	-13,363	-	-	-	-	-	-
150												
151	Geared Cash Flow		-253,888	-	-	-	16,279	16,279	16,279	16,279	-20,991	800,152
152												
153	ICR			3.42	3.50	3.59						
154	DSCR			1.00	1.00	1.00						
155	LTV			46.25%	45.15%	44.04%						
156	Debt Yield			13.51%	13.84%	14.19%						

Figure 7.6 Formatting.

Covenants and Geared Returns

We have so far calculated the returns for the I/O structure only, but we still have another five to go. It sounds tedious… A way around it is to calculate the 'live' returns and create a data table to automatically calculate covenants and geared returns for each structure.

Calculate 'Live' Covenants and Returns

For the live target return, use the INDEX function. The INDEX function returns the value at a given position in a range or array and its syntax is:

= INDEX (array, row number, column number)

In our case, it will be:

> Target Return = **INDEX** (target returns per structure, debt structure number)

Because we are indexing only one row, we do not need to enter the row number, there is only one. Therefore, you will need to add *two* commas to the function (it wasn't a typo!).

	B	C
25	**Covenants Output**	
26	Min ICR	=MIN(O153:BC153)
27	Min DSCR	=MIN(O154:BC154)
28	Max LTV	=MAX(O155:BC155)
29	Min Debt Yield	=MIN(O156:BC156)
30		
31		
32	**Structures Output**	
33	*Target Return*	=INDEX(D33:I33,,N138)
34	Debt No.	
35	IRR	=XIRR(O151:BC151, O5:BC5)
36	NPV	=XNPV(C33,O151:BC151,O5:BC5)
37	Worth	=C36+gross_price
38	EAR	=XIRR(O149:BC149,O5:BC5)

Figure 7.7 Covenants 1.

Create a Data Table

Highlight the table for the *covenants output* first:

	B	C	D	E	F	G	H	I
24								
25	Covenants Output		1	2	3	4	5	6
26	Min ICR	3.42						
27	Min DSCR	1.00						
28	Max LTV	46%						
29	Min Debt Yield	0%						

Figure 7.8 Covenants 2.

Go to **Data > What If Analysis > Data Table**

Figure 7.9 Covenants 3.

In **Row input cell,** select the debt structure number:

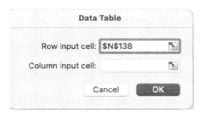

Figure 7.10 Covenants 4.

Click **OK.**

	B	C	D	E	F	G	H	I
24								
25	Covenants Output		1	2	3	4	5	6
26	Min ICR	3.42	3.86x	3.33x	3.41x	3.32x	2.66x	3.42x
27	Min DSCR	1.00	3.86x	2.22x	1.08x	2.30x	2.66x	1.00x
28	Max LTV	46%	51%	48.96%	46.38%	49.15%	65.41%	46.25%
29	Min Debt Yield	0%	13%	13.27%	13.48%	13.26%	10.70%	0.00%

Figure 7.11 Covenants 5.

Do the same for the Structures Output (Highlight > Data Table).

	A	B	C	D	E	F	G	H	I
32		Structures Output							
33		Target Return	15.00%	15.00%	15.00%	15.00%	15.00%	15.00%	15.00%
34		Debt No.		1	2	3	4	5	6
35		IRR	12.90%						
36		NPV	-42,388						
37		Worth	625,737						
38		EAR	4.31%						

Figure 7.12 Covenants 6.

	A	B	C	D	E	F	G	H	I
32		Structures Output							
33		Target Return	15.00%	15.00%	15.00%	15.00%	15.00%	15.00%	15.00%
34		Debt No.		1	2	3	4	5	6
35		IRR	12.90%	18.35%	16.97%	13.43%	16.99%	22.42%	12.90%
36		NPV	-42,388	45,053	28,780	-30,285	29,043	81,883	-42,388
37		Worth	625,737	713,178	696,905	637,840	697,168	750,008	625,737
38		EAR	4.31%	4.18%	4.20%	4.28%	4.20%	4.16%	4.31%

Figure 7.13 Covenants 7.

Finding the Maximum LTV

The idea is to find the maximum borrowing amount without breaching any covenants, and for this, we will use the **SOLVER** function in Excel.

EXERCISE 7

An investor is considering borrowing on an interest-only debt structure and were given a set of covenants (ICR, DSCR, LTV and Debt Yield) that they need to comply to. Given, the forecast table for the property below, what is the maximum LTV that the investor can borrow at?

First, let's look at the forecast table to:

	P	Q	R	S
6				
7	Property Forecast			
8	Year	Rental Growth		Initial Yield
9	Year	% p.a.	% p.q.	
10	2021	-2.00%	-0.50%	5.75%
11	2022	0.00%	0.00%	6.06%
12	2023	-1.00%	-0.25%	6.38%
13	2024	1.50%	0.37%	6.69%
14	2025	2.00%	0.50%	7.00%
15				
16	Source:	Example Research		
17	Date:	Nov-21		

Figure 7.14 Maximum LTV 1.

Then, the LTV covenant is currently set at 60% but this is the input we will need to change in order to find the maximum LTV:

	A	B	C	D	E	F	G	H	I
17									
18		Covenants							
19		ICR		2.00x	2.00x	2.00x	2.00x	2.00x	2.00x
20		DSCR		1.50x	1.50x	1.50x	1.50x	1.50x	1.50x
21		LTV		60.0%	70.0%	70.0%	70.0%	70.0%	70.0%
22		Debt Yield		5.00%	5.00%	5.00%	5.00%	5.00%	5.00%

Figure 7.15 Maximum LTV 2.

To find the maximum LTV, go to **Data > Solver:**

Set Objective >> LTV **To MAX By Changing Variable Cells** >> LTV
Add Constraints:
ICR Covenant <= ICR Calculated
DSCR Covenant <= DSCR Calculated
LTV Covenant >= LTV Calculated
Debt Yield Covenant <= Debt Yield Calculated

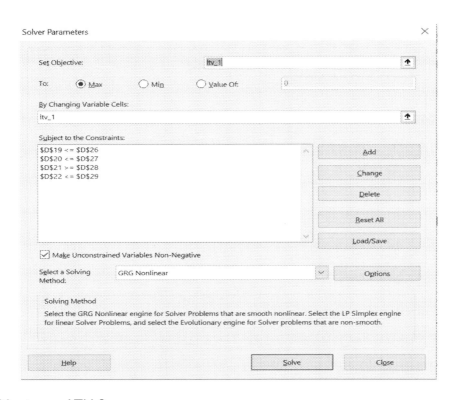

Figure 7.16 Maximum LTV 3.

We find that the maximum LTV is now 55%; otherwise, the LTV would be breached during the life of the loan.

	A	B	C	D
2		**DEBT DETAILS**		
3				
4		Debt No.		1
5		Type		I/O
6		Interest rate type		Fixed
7		LTV		54.66%
8		Fixed - Int. Rate		4.00%
9		Floating - Margin		
10		Amortisation		
11		Amort. Period		
12		Maturity		10 yrs
13				
14		Fees		
15		Origination fee		1.00%
16		Rolled fee?		Yes
17				
18		Covenants		
19		ICR		2.00x
20		DSCR		1.50x
21		LTV		60.0%
22		Debt Yield		5.00%
23				
24				
25		Covenants Output		1
26		Min ICR		4.16x
27		Min DSCR		4.16x
28		Max LTV		60.00%
29		Min Debt Yield		16.67%
30				
31				
32		Structures Output		
33		Target Return		15.00%
34		Debt No.		1
35		IRR		21.38%
36		NPV		94,336
37		Worth		694,336
38		EAR		4.18%

Figure 7.17 Maximum LTV 4.

Geared vs. Ungeared Returns

Given the structures discussed so far and assuming inputs are correct for all loans, which debt structure would be most suitable? Why?

EXERCISE 9

Create an output table showing geared and ungeared returns, as per Asset sheet output. Create a sensitivity analysis based on exit yields.

Linking Chosen to Live Returns

For us to calculate the geared returns, we will need to make a decision about which senior debt structure we will choose. We will choose that with highest project IRR, NPV, worth, etc. Note that in our case, all interest rates, covenants, and terms were the same, but it's not always so; in reality, it is very rarely so as it wouldn't make economic sense, since some structures are riskier than others from the lender's perspective.

As we were looking at the floating – interest-only structure, we will choose this one and make a system to link it to the 'Live' cash flow.

Comparing Debt Structures and Analysing Results 157

On the **Asset** sheet, go to **Data** and create a **Data Validation** with the debt types:

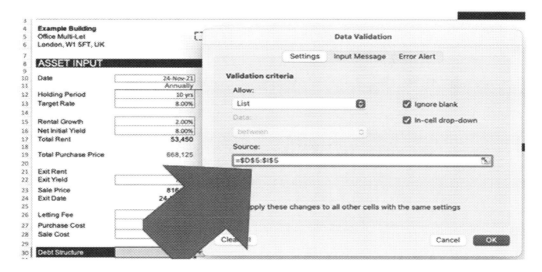

Figure 7.18 Linking Chosen to Live Returns.

Then, link onto the cell next to it, the Debt Number, using the **INDEX MATCH** function:

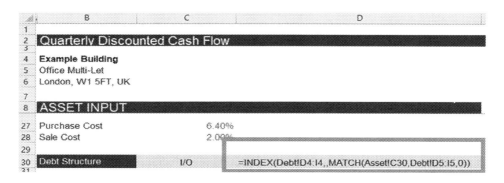

Figure 7.19 Linking Chosen to Live Returns 2.

Then, link the 'Live' cash flow to the number of the INDEX MATCH function.

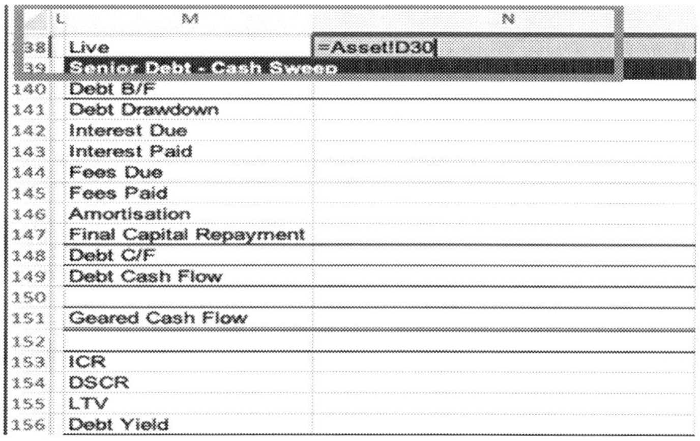

Figure 7.20 10.3 Linking Chosen to Live Returns 3.

The Output Table

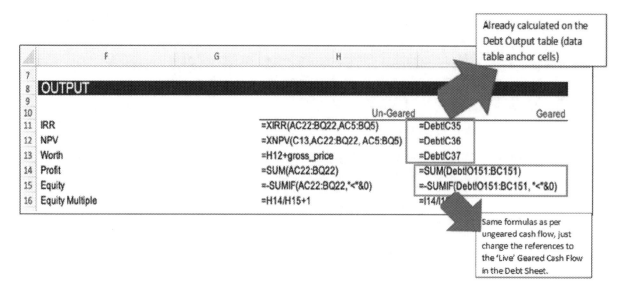

Figure 7.21 Linking Chosen to Live Returns 4.

	F	G	H	I
7				
8	**OUTPUT**			
9				
10			Un-Geared	Geared
11	IRR		9.93%	13.43%
12	NPV		87,255.67	-30,285
13	Worth		755,380.67	637,840
14	Profit		674,354	581,990
15	Equity		675,354	286,810
16	Equity Multiple		2.00	3.03

Figure 7.22 Linking Chosen to Live Returns 5.

Sensitivity Analysis

This will show you how your outputs will change when inputs are varied. In the example below, we will see the impact of exit yields varying from 6% to 8.5% in 50bps on outputs.

Steps:

1 Create a table with exit yields in a row and outputs in a column.
2 Then, you will need to link the calculation cells of each output; these will be the 'anchor' cells.
3 The outputs (anchor cells) will need to be shown on the cell to the left and below the first exit yield.
4 Highlight the relevant cells, i.e., the array/table between the exit yields and the anchor cells.
5 Go to **Data** > **What IF Analysis** > **Data Table**.
6 Then, in **Row Input Cell,** enter the cell for the exit yield.
7 Select **OK**.

The table will be populated with the values:

	F	G	H	I	J	K	L	M
18	**Sensitivity**							
19	Exit Yield		0.06	0.065	0.07	0.075	0.08	0.085
20	=F11	=I11	=TABLE(C22,)	=TABLE(C22,)	=TABLE(C22,)	=TABLE(C22,)	=TABLE(C22,)	=TABLE(C22,)
21	=F12	=I12	=TABLE(C22,)	=TABLE(C22,)	=TABLE(C22,)	=TABLE(C22,)	=TABLE(C22,)	=TABLE(C22,)
22	=F13	=I13	=TABLE(C22,)	=TABLE(C22,)	=TABLE(C22,)	=TABLE(C22,)	=TABLE(C22,)	=TABLE(C22,)
23	=F14	=I14	=TABLE(C22,)	=TABLE(C22,)	=TABLE(C22,)	=TABLE(C22,)	=TABLE(C22,)	=TABLE(C22,)
24	=F15	=I15	=TABLE(C22,)	=TABLE(C22,)	=TABLE(C22,)	=TABLE(C22,)	=TABLE(C22,)	=TABLE(C22,)
25	=F16	=I16	=TABLE(C22,)	=TABLE(C22,)	=TABLE(C22,)	=TABLE(C22,)	=TABLE(C22,)	=TABLE(C22,)

Figure 7.23 Sensitivity Analysis.

Credit Enhancements

Credit enhancements are instruments that reduce the credit risk to the lenders and thus may lower the funding cost (interest rate or margin) to the borrower or indeed may allow a debt to be approved that would not be approved otherwise. The nature and the amount of credit enhancements are determined by the lending house: banks, insurance and pension funds, mezzanine providers, and debt funds and can be internal or external.

In this course, we will look at a debt service reserve account – or cash reserve – as an internal credit enhancement facility and a revolving loan as the externally provided facility.

Debt Service Reserve Account (Cash Reserve)

The idea is that a cash reserve will ensure that there is enough liquidity in the project to meet the debt service requirements (interest and principal payments). A typical requirement is that the borrower puts a certain amount of cash aside, normally six months of prospective debt service.

This account ensures that if the project has a negative cash flow for the reasons of capital expenditures, vacancy, and operational expenses, the debt will still be paid. Another feature is that if something bad happens and the debt needs to be restructured, there is sufficient time for the borrower to make alternative arrangements.

The problem with a cash reserve is that holding cash can have a negative effect on the overall project IRR and NPV, since the return on cash will be significantly lower than the project returns (the IRR). Say, if a project has an estimated IRR of 7% and the income received in the cash account is 1%, this rate differential can lower the equity IRR rather considerably.

The amount used from the cash reserve derives from the geared cash flow after the senior debt. Therefore, this can be considered a subordinated account and form a cash flow waterfall.

Cash Reserve Modelling

Principal:

This is the initial amount that the borrower will need to provide as a reserve in the account. Lenders may calculate this amount based on the debt service requirements over a period of six months or it may be a percentage of the total loan amount.

Min. Balance Required:

This is the minimum balance required in the account. For example, if we say that the principal is £3 million and the minimum is £2 million, the cash reserve can use up to £1 million to cover debt services before the lender 'calls' for new money into the account.

Interest Income:

The interest income will be based on current cash accounts in the market. Typically, the interest rate received on cash accounts are minimal and this is what is normally termed 'cash drag'.

Min ICR:

The minimum ICR will come into question when it is breached by the project cash flow after the debt service (geared cash flow). The main reason is that the lenders will then require that excess cash flow is retained in the cash reserve until a certain time that the ICR is at or above the minimum ICR. This situation is called **'cash trap'**.

Min DSCR:

This follows the same rationale and mechanics as for the minimum ICR.

Cell Names:

	B	C	D
49	**Reserve and Revolver**		
50		Cash	Revolving
51		Reserve	Facility
52	Principal	res_p	rev_p
53	Min. Balance Required	min_res_req	
54	Interest rate	res_rate	rev_Rate
55			
56	Min ICR	min_res_ICR	
57	Min DSCR	min_res_dscr	
58			
59	Commitment Fee		com_fee
60	Utlisation Fee		util_fee
61	Upfront Fee		rev_upfee

Figure 7.24 Cell Names.

Layout:

	L	M	N	O	P	Q
157						
158		**Cash Reserve**				
159		Reserve B/F				
160		Cash Inflow				
161		Cash Outflow				
162		Calls				
163		Withdrawals				
164		Reserve C/F				
165		Reserve Cash Flow				
166						
167		Interest Income				
168		Cash Flow after Reserve				

Figure 7.25 Layout.

Labels & Formulas:

RESERVE B/F

The opening balance of the reserve account.

CASH INFLOW

In Period 0, the inflow is the amount of capital we will need to set up in the cash reserve. Then, if any of covenants ICR or DSCR are breached, then there will be a cash sweep until the covenants are brought back into compliance. This is called **Cash Trap**.

CASH OUTFLOW

Conversely, the reserve account will see a cash outflow if the geared cash flow is negative. In this case, the borrower will take money out of the cash reserve only if there hasn't been a breach in terms of minimum ICR or DSCR and the geared cash flow is negative. This is called **Cash Release**.

CALLS

Calls refer to the time when cash will be 'called' into the account if the cash balance is below the minimum amount required.

WITHDRAWALS

Cash can be taken out from the reserve when the covenants pass and the amount held in the account is above the initial principal. Also, you may withdraw the cash as the loan term expires.

INTEREST INCOME

Because the cash reserve will be sitting in a bank account, it should attract interest income.

NET CASH FLOW

> = Geared Cash Flow + Reserve Cash Flow + Interest Income

RESERVE C/F

= Sum of everything above.

RESERVE CASH FLOW

= - SUM (Cash Inflow – Cash Outflow + Income Interest +/- Calls or Withdrawals)

	M	N	O
158	**Cash Reserve**		
159	Reserve B/F	=N164	
160	Cash Inflow	=IF(O4=0,res_p,(O140>0) * OR((O153<min_res_ICR), (O154<min_res_dscr)) * MAX(0,O151))	
161	Cash Outflow	=-(O140>0) * (O153>min_res_ICR)*(O154>min_res_dscr) * (MIN(- MIN(0,O151), min_res_req))	
162	Calls	=(O140>0) * (SUM(O159:O161)<min_res_req) * (res_p-SUM(O159:O161))	
163	Withdrawals	=-IF(O148=0, SUM(O159:O162), (O153>min_res_ICR) * (O154>min_res_dscr) * (SUM(O159:O161)>res_p) * (SUM(O159:O161)-res_p))	
164	Reserve C/F	=SUM(O159:O163)	
165	Reserve Cash Flow	=-SUM(O160:O163)	
166			
167	Interest Income	=O159*res_rate*YEARFRAC(N5, O5, 1)	
168	Cash Flow after Reserve	=O151+O165+O167	

Figure 7.26 Reserve Cash Flow.

Revolving Facility Modelling

Revolving debt facilities (revolving loan) is also used as credit enhancement. The idea is similar to debt service reserve account; however, as an external credit facility, there will be interest to be paid on the amount available as credit (commitment fee) and the amount actually used (utilisation fee).

	M	N	O	P	Q	R	S	T	U	BA	BB	BC
158	**Cash Reserve**											
159	Reserve B/F		-	500,000	500,000	500,000	500,000	500,000	500,000	-	-	-
160	Cash Inflow		500,000	-	-	-	-	-	-	-	-	-
161	Cash Outflow		-	-	-	-	-	-	-	-	-	-
162	Calls		-	-	-	-	-	-	-	-	-	-
163	Withdrawals		-	-	-	-	-	-	-	-	-	-
164	Reserve C/F		500,000	500,000	500,000	500,000	500,000	500,000	500,000	-	-	-
165	Reserve Cash Flow		-500,000	-	-	-	-	-	-	-	-	-
166												
167	Interest Income		-	4,932	4,986	5,041	5,041	4,932	4,986	-	-	-
168	Cash Flow after Reserve		-753,888	4,932	4,986	5,041	5,041	4,932	4,986	16,279	-20,591	800,152

Figure 7.27 Reserve Cash Flow 2.

Upfront Fee

This is also called front-end fees and are paid on signing or by first utilisation.

They are paid on the whole of the facility regardless of whether the facility is subsequently under-utilised, cancelled, or pre-paid.

Commitment Fee

These are generally charged by the lender during the revolving period for the debt that is committed but not yet used, i.e., the undrawn balance.

Utilisation Fee

They are based on the amount of credit used by the borrower in the revolving loan.

Expiry

This is when the revolving debt facility expires. In our case, we will assume it's an 'evergreen' loan, i.e., without expiry date.

Layout

	L	M	N	O	P
169					
170		**Revolving Facility**			
171		Credit B/F			
172		Loan Drawdown			
173		Principal Repayment			
174		Interest Due			
175		Upfront Fees			
176		Commitment Fees			
177		Utilisation Fees			
178		Interest + Fees Paid			
179		Credit C/F			
180		Revolving facility cash			

Figure 7.28 Layout.

Loan Drawdown

As part of the 'waterfall', the borrower will only draw from the revolving debt if there is a negative cash flow that needs to be covered.

Principal Repayment

This is the amount repaid from the revolving facility. It assumes that if the cash from the project is positive, the excess amount will be used to repay the debt.

Interest + Fees Paid

We will assume that all fees are paid in its entirety.

	M	N	O
170	**Revolving Facility**		
171	Credit B/F		=N179
172	Loan Drawdown		=(O140>0) * (O168<0) * (O171<rev_p) * MIN(-MIN(O168,0), res_p-O171-SUM(O174:O177))
173	Principal Repayment		=IF((O4>hold_period), 0, -IF(O148=0, O171, IF(O171+SUM(O174:O178)>rev_p, SUM(O174:O178) + O171-rev_p, MIN(MAX(O168-SUM(O174:O177),0), O171))))
174	Interest Due		=IF(OR(O4>hold_period, O4=0), 0, rev_rate * YEARFRAC(N5, O5) * O171)
175	Upfront Fees		=(O4=0) * rev_upFee * rev_p
176	Commitment Fees		=(O4>0) * com_fee * YEARFRAC(N5,O5) * (rev_p-O171)
177	Utilisation Fees		=IF(O4>hold_period, 0, util_fee * YEARFRAC(N5,O5) * O171)
178	Interest + Fees Paid		=-MIN(MAX(0,O168), SUM(O174:O177))
179	Credit C/F		=SUM(O171:O178)
180	Revolving facility Cash Fl		=O172+O173+O178
181			
182	Cash Flow After Revolver		=O151+O180

Figure 7.29 Interest + Fees Paid.

	M	N	O	P	Q	R	S	T	U	BA	BB	BC
170	**Revolving Facility**											
171	Credit B/F		-	15,000	14,643	14,230	13,760	13,288	12,923	-	-	4,500
172	Loan Drawdown		-	-	-	-	-	-	-	-	-	-
173	Principal Repayment		-	-357	-413	-470	-472	-365	-422	-	-	-4,500
174	Interest Due		-	113	110	107	103	100	97	-	-	34
175	Upfront Fees	15,000	-	-	-	-	-	-	-	-	-	-
176	Commitment Fees		-	4,388	4,390	4,393	4,397	4,400	4,403	4,500	4,500	4,466
177	Utilisation Fees		-	75	73	71	69	66	65	-	-	23
178	Interest + Fees Paid		-	-4,575	-4,573	-4,571	-4,569	-4,566	-4,565	-4,500	-	-4,523
179	Credit C/F	15,000		14,643	14,230	13,760	13,288	12,923	12,501	-	4,500	-
180	Revolving facility Cash Flow		-	-4,932	-4,986	-5,041	-5,041	-4,932	-4,986	-4,500	-	-9,023
181												
182	Cash Flow After Revolver		-253,888	-4,932	-4,986	-5,041	-5,041	-4,932	-4,986	11,779	-20,591	791,130

Figure 7.30 Interest + Fees Paid 2.

Conclusion

In this chapter, we created a system to analyse the geared investment by developing a 'live' debt schedule that will dynamically calculate the outputs, such as geared IRR, NPV, equity multiple, and the covenants; this way, we can highlight any breaches and adjust the inputs accordingly, such as the loan to value, using the Goal Seek function in Excel.

In this chapter, we also looked at credit enhancement techniques – and how to model them – such as the cash reserve and revolving facilities.

By the end of this chapter, you should be able to analyse different debt structures and how they can gear a real estate investment by means of a senior debt facility.

Section 3

Real Estate Development Financial Modelling

We will now start creating the development financial model, and because it is important that you know both how to model this and the concepts, I will give some brief explanations of terminology used in real estate development and walk you through the inputs and definitions as well.

Chapter 8

Development Valuation and Analysis

Chapter Contents
Introduction 166
Development Time 167
    Time as a Critical Risk Element 167
    Development Phases and Durations Input 168
    Construction Period and Development Status 172
    Development Costs Distribution 173
    Pro-Forma Development Cash Flow 176
    Residual Value on IRR 181
    Goal Seek vs Solver 182
Conclusion 183

Introduction

Development valuation is generally the term used to assess the revenue from a property not yet built and therefore the estimated value of the completed building or the gross and net development values (GDV and NDV).

However, as developers are not only interested in finding out the development value, but the potential profit resulting from this endeavour, further calculations need to be done to include development costs, such as construction, professional fees, financing costs, and the site costs. As such, further to the development valuation, surveyors can run a development appraisal and the residual valuation.

Development appraisal is the process of finding the profit of the development – both in terms of profit on cost and the net present value (NPV) and internal rate of return (IRR). This process can also be referred to as the viability analysis. In this case, the valuer will come up with a profit figure after deducting the development costs, such as for infrastructure, construction, professional fees, planning obligations, financing, and site or land costs. If this profit – either the undiscounted profit or discounted, resulting in an NPV – is enough, then it would be worth the developer's time and risk, and in such case, we would say that the development is viable.

Residual valuation is then the process of finding the site or land value, or the site's maximum bid price given the assumptions in the development appraisal. The site value is then a function of the required profit that the developer needs to make for the project to be viable, and hence, the word 'residual', because the site or land becomes the residual ('leftover') value after all the considerations for revenue, costs, and profit were taken into account.

Another term commonly used is development analysis. This is a generic term which groups a variety of pre-construction studies by generalists, for example, surveyors, and specialists, for example, structural engineers, in a systematic method of inquiry to determine facts that should be reliable assumptions about the future of the property scheme. In our 'world' of real estate investment and finance, we assume that specialist construction professionals have already analysed the proposed scheme and we are happy with their findings or at least convinced that these are factual. Our job is therefore to estimate the economic and financial aspects of the development, i.e., its financial risk and return. Development analysis will therefore incorporate risk analysis techniques, such as sensitivity analysis, scenario analysis, and simulation for example.

In summary:

Development valuation: Where the valuer will determine the gross and net development value of the scheme: GDV and NDV, respectively.

Residual valuation: Where the valuer will determine what should be paid for the site and a profit is targeted.

Development appraisal: Where the valuer will determine the profitability of a real estate project, either as an NPV or undiscounted profit, and determine whether the development project is financially and economically viable.

Development analysis: Where the valuer will use risk analysis techniques to identify the impact of changes in input on the outcome, this outcome being the development value, residual value, or profit.

Development valuations, development appraisals, and residual valuations can be carried out on a short-cut, i.e., all-risk yield and profit method or on a discounted cash flow (DCF) method. The DCF method will provide a more explicit analysis of the scheme and allow you to estimate the internal rate of return (IRR) and net present value (NPV), which can then be sensitised for the effects of time delays and economic changes for the development analysis.

Development Time

Time as a Critical Risk Element

Time delay is a fundamental risk in the development process. Time permits the power of compound interest to erode the developer's resources, and it allows the conditions of competition and consumer needs which were true when the project started to change significantly. Cost inflation especially in the construction industry will also erode returns, especially in times when labour and commodities costs are high.

In terms of finance charges, a project with £1,000,000 invested at a nominal construction interest rate as low as 6% per annum is costing £5,000 interest for the first month, £167 a day, and then £5,600 the second month and so on. If the developer had hoped for a net profit of £25,000, a total delay of 4 months in completing the project will not only cause the loss of that profit in additional interest charges but also may give the tenant the right to break their lease, the owner the right to invoke a loss of use penalty, the mortgage lender the right to renegotiate more expensive terms than those in the original commitment and a competitor the opportunity to finish first and capture the market.

As money and time are expended on the project, time becomes of the essence in achieving expected revenues from sales and rentals. Delays in construction also mean poorer IRRs and NPVs as costs are recognised earlier whereas revenues later.

Time also adds to the uncertainty of construction costs; the actual costs will exceed any contingency allowance, and the longer the project takes, the higher the risks related to completion.

The total development time needs to allow for obtaining planning consent, preparing drawings, and so on. This is referred to as lead-in time because it precedes the construction phase. If the development site is an existing building with tenants in place, then this may add to the lead-in period as well.

There may also be a period of time between completion of the shell and core leaving the fit-outs, or tenancy improvement, to be completed when the building is able to be occupied.

Development Timeline:

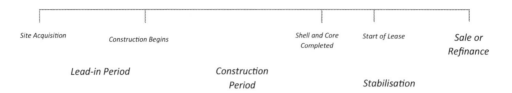

Figure 8.1 Development Phases.

Development Phases and Durations Input

	B	C D	E	F	G	H
4	**SITE**					
5	Project Name		Cambridge Finance Development Scheme 1			
6	Address		Project 1 Address			
7	Date		17-Oct-22			
11	**TIMESCALE**					
12			Status No	Start Date	Duration	End Date
13	Planning, Design, Pre Build		1	31-Dec-22	3 qtrs	30-Sep-23
14	Construction		2	30-Sep-23	8 qtrs	30-Sep-25
15	Vacancy		3	30-Sep-25	2 qtrs	31-Mar-26
16	Stabilisation		4	31-Mar-26	4 qtrs	31-Mar-27
17	Sale (Exit)		0	31-Mar-27		

Figure 8.2 Development Timeline 2.

	B	C E	F	G	H
4	**SITE**				
5	Project Name		Cambridge Finance Development Scheme 1		
6	Address		Project 1 Address		
7	Date		44851		
11	**TIMESCALE**				
12		Status No	Start Date	Duration	End Date
13	Planning, Design, Pre Build	1	=DATE(YEAR(E7),ROUNDUP(MONTH(E7)/3,0)*3+1, 0)	3	=EOMONTH(F13,G13*3)
14	Construction	2	=H13	8	=EOMONTH(const_start,const_dur*3)
15	Vacancy	3	=const_end	2	=EOMONTH(F15,G15*3)
16	Stabilisation	4	=H15	4	=EOMONTH(F16,G16*3)
17	Sale (Exit)	0	=H16		

Figure 8.3 Development Timeline 3.

Gross Development Value (GDV) vs Net Development Value (NDV) vs Sales Proceeds

In order to be consistent with the market approach of gross and net values, I will use GDV – gross development value – as the total value of the finished property which *includes* purchaser's costs. Therefore, the calculation for GDV is defined as:

Development Valuation and Analysis 169

> GDV = Net Rent/Net Yield

To calculate the developer's income, we need to arrive at the **net development value (NDV)**. To do so, we deduct the purchaser's costs from the GDV. If the purchaser's costs assumed in the net yield is say 6.5%, then to arrive at the NDV, we need to divide the GDV by (1 + purchaser's costs) or (1 + 6.5%).

> NDV = GDV/(1 + purchaser's costs)

Sometimes, even the NDV is not the net sales amount that the developer will achieve since there may be some concessions, for example, a rent-free period top-up, a fit-out contribution for the units still to be leased, or some rent guarantees. Moreover, the developer will need to consider the sales costs, for example, brokers and legal costs.

> Sales Proceeds = NDV − Concessions − Sales Costs

Development Revenue Inputs

	A	B	C	D	E	F	G	H
19		VALUATION						
20								
21		REVENUES			NIA	Rent	Net Yield	Gross
22					ft2	ft2	%	Value
23		Office Units			5,400	50.00	4.00%	6,750,000
24		GROSS DEVELOPMENT VALUE (GDV)						6,750,000
25								
26		Purchaser's Cost					6.40%	
27		NET DEVELOPMENT VALUE (NDV)						6,343,985
28								
29		Concessions (Rent Free, Capex contribution, Stamp duty)					4.00%	253,759
30		Sales Costs					2.50%	158,600
31		**Sales Proceeds**						**5,931,626**

Figure 8.4 Development Revenue Inputs.

	A	B	C	D	E	F	G	H
19		VALUATION						
20								
21		REVENUES			NIA	Rent	Net Yield	Gross
22					ft2	ft2	%	Value
23		Office Units			5400	50	0.04	=(E23*F23)/G23
24		GROSS DEVELOPMENT VALUE (GDV)						=SUM(H23:H23)
25								
26		Purchaser's Cost					0.064	
27		NET DEVELOPMENT VALUE (NDV)						=H24/(1+G26)
28								
29		Concessions (Rent Free, Capex contribution, Stamp duty)					0.04	=H27*G29
30		Sales Costs					0.025	=G30*H27
31		**Sales Proceeds**						**=H27-SUM(H29:H30)**

Figure 8.5 Development Revenue Inputs 2.

Development Costs

Construction costs and fees can be more specifically estimated by a quantity surveyor, but an approximation can be gained by reference to recent contracts for similar developments or from building price books such as Spon's Architects and Builders Price Book or BCIS.

Development costs are made up of site costs, hard costs, soft costs and finance costs.

COSTS

Site Costs

Site costs refer to the purchase price of the site. If the developer is buying bare land, then it is the land price. If the site is on the market, then we should be the asking price and calculated as the total site costs after purchaser's costs (stamp duty, etc.). If there needs to be a residual valuation calculation to determine the site costs, then initially just add any hardcoded figure, or even zero, so we can then calculate the residual site or land value once we have finalised the appraisal and set a target profit, either through profit on cost or IRR.

Another important feature of the site costs, other than becoming the residual value, is that it is possible that the developers only pay either a small percentage (around 10% typically) as a deposit or an option at the start of the development but the remaining value is then paid once planning permission is granted or finance obtained and this is normally assumed to be at the start of the construction period.

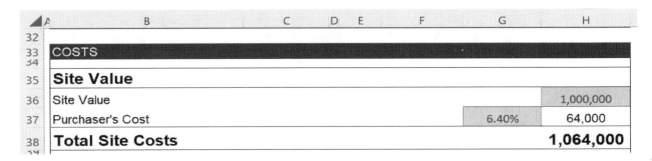

Figure 8.6 Site Costs.

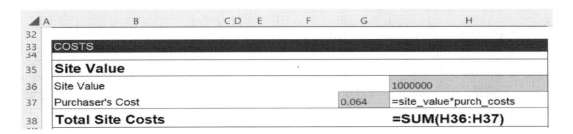

Figure 8.7 Site Costs 2.

Hard Costs

Hard costs refer to those costs associated with the physical construction of the building and can be related to the building's structure, the site, and to the landscape. All labour and materials required for construction are included in hard costs. In terms of the building site, all utilities, life safety systems, and equipment, HVAC systems, paving, grading, etc. are considered hard costs. Hard costs associated with the landscape are based on the architectural drawings and include grass, trees, mulch, fertiliser, etc.

Generally, hard costs are more tangible and therefore easier to estimate. The range of hard costs varies widely but tend to be most expensive where labour is more expensive, for example, Central London.

	B	C	D	E	F	G	H
40	**HARD COSTS**						
41	CONSTRUCTION COSTS				GIA		
42	Construction			Eff.	ft2	Rate ft2	Total Cost
43	Office Units			90%	6,000	300.00	1,800,000
44	Total						1,800,000
45							
46	**CONTINGENCY**					10%	180,000
47							
48	SITE INVESTIGATION, PREPARATION AND INFRASTRUCTURE COSTS						
49	Site Investigation						50,000
50	Demolition						45,000
51	Landscaping						50,000
52	Total						145,000
54	**TOTAL HARD COSTS**						2,125,000
55	Distribution Curve						S-Curve

Figure 8.8 Hard Costs 1.

	B	C	D	E	F	G	H
40	**HARD COSTS**						
41	CONSTRUCTION COSTS				GIA		
42	Construction			Eff.	ft2	Rate ft2	Total Cost
43	=B23			0.9	=E23/E43	300	=G43*F43
44	Total						=SUM(H43:H43)
45							
46	**CONTINGENCY**					0.1	=G46*H44
47							
48	SITE INVESTIGATION, PREPARATION AND INFRASTRUCTURE COSTS						
49	Site Investigation						50000
50	Demolition						45000
51	Landscaping						50000
52	Total						=SUM(H49:H51)
54	**TOTAL HARD COSTS**						=H44+H46+H52
55	Distribution Curve						S-Curve

Figure 8.9 Hard Costs 2.

Soft Costs
Soft costs are any costs that are not considered direct construction costs. Soft costs include everything from architectural and engineering fees to legal fees, pre- and post-construction expenses, permits and taxes, insurance, planning obligations, etc. Soft costs also include movable furniture and equipment (as opposed to fixed equipment included in hard costs) such as computer data equipment, telephone systems, etc. Depending on the project, soft costs can also include expenses that continue after completion such as building maintenance, insurance, security, and other fees associated with the asset's upkeep.

Sales costs can also be included in soft costs, such as marketing, legal fees, and agent fees even though they occur after the development has been completed. However, we have included sales costs in the sales proceeds to make a direct relationship between sales costs and achieved revenue.

	A	B	C	D	E	F	G	H
58		**SOFT COSTS**						
59		**PROFESSIONAL FEES**					% of HC	
60		Planning					1.00%	21,250
61		Architect					10.00%	212,500
62		All Other Prof					5.00%	106,250
63		Total						**340,000**
64								
65		**PLANNING OBLIGATIONS**						
66		CIL						100,000
67		S106						100,000
68		Other Obligations						5,000
69		Total						**205,000**
72		**TOTAL SOFT COSTS**						**545,000**
73		Distribution Curve						Straight Line
76		**TOTAL COSTS EX. FINANCING**						**3,734,000**

Figure 8.10 Soft Costs 1.

	A	B	C	D	E	F	G	H
58		**SOFT COSTS**						
59		**PROFESSIONAL FEES**					% of HC	
60		Planning					0.01	=G60*hard_costs
61		Architect					0.1	=G61*hard_costs
62		All Other Prof					0.05	=G62*hard_costs
63		Total						=SUM(H60:H62)
65		**PLANNING OBLIGATIONS**						
66		CIL						100000
67		S106						100000
68		Other Obligations						5000
69		Total						=SUM(H66:H68)
72		**TOTAL SOFT COSTS**						=H63+H69
73		Distribution Curve						Straight Line
76		**TOTAL COSTS EX. FINANCING**						=total_site+hard_costs+soft_costs

Figure 8.11 Soft Costs 2.

Construction Period and Development Status

To distribute the cash inflows and outflows, we first need to create the periods, dates, and then some intermediate calculations, such as development statuses and construction periods.

The development statuses refer to the timeline as seen in the beginning of the model, and we use a LOOKUP function of the dates and the development timeline.

As the construction period is the date after the construction start and before the construction end, we need to add one to the previous period; otherwise, it should be zero.

	A	B	C	D	E	F	G	H	I	J
80		**COST CURVES**								
83		Period					0	1	2	3	4
84		Date					31-Dec-22	31-Mar-23	30-Jun-23	30-Sep-23	31-Dec
86		Development Status					1	1	1	2	2
87		Construction Period					0	0	0	1	2

Figure 8.12 Construction Period 1.

Development Valuation and Analysis 173

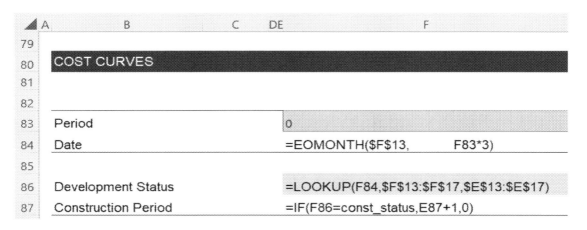

Figure 8.13 Construction Period 2.

Development Costs Distribution

If you have already involved contractors and quantity surveyors, they will give you the amount that you will need to disburse month on month, quarter on quarter, and so on. However, as we do the first analysis of the feasibility of the development, we will need to make approximations to have a view of how the cash flow will look like. These are called the cost curves, and they can be:

- Straight line
- S-curve
- Known curve

Straight Line

Straight line cost distribution means that during the period of distribution, the cost will be distributed equally period by period.

In our assumption, we will distribute soft costs as a straight-line curve and over the period of the whole development:

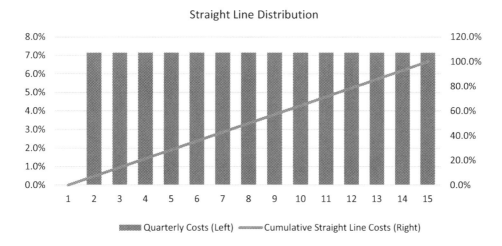

Figure 8.14 Straight Line 1.

	A	B	C	D	E	F	G	H	I
88			Check						
89		Soft Cost Distribution	0.0%			0.0%	33.3%	33.3%	33.3%

Figure 8.15 Straight Line 2.

	A	B	C	D	E	F
88			Check			
89		Soft Cost Distribution	=1-SUM(F89:Z89)			=IF(E86=pre_build,1/plan_dur,0)

Figure 8.16 Straight Line 3.

Note that I have also added an **_Error Check_**. We want the total distribution to add to one (100%), but if the sum of all percentages is different than one, then Excel will show us the discrepancy and we can check the formula or assumptions for error.

S-Curve

We will use the S-curve to distribute the hard costs, and as such, this curve assumes that the costs are not very much in the beginning, but will gradually increase to a peak – generally at the middle of construction period – then it will tail off again. We normally refer to this type of cost distribution as the S-curve, noting that the S-curve is one that assumes that the costs are distributed following a normal distribution.

We will first need to create intermediate calculations using the normal distribution curve: Z-score, s-curve, and draws.

Z-SCORE

To understand the S-curve, we will need to use statistical assumptions. The S-curve indicates that the construction costs will use a normal distribution, which has six standard deviations in total. This means that the total number of construction periods will be within 6 standard deviations, 3 to the left and 3 to the right of the mean. For example, if the total duration of construction is 2 years, or 24 months or 8 quarters, it will mean that the standard deviation will be either =24 months/6 or 4 months; or 8 quarters/6 or 1.33 quarters.

S-CURVE

This is the cumulative normal curve by period.

The =**NORMSDIST()** function provides the cumulative percentage from the left of a standard normal curve value.

Note: Excel in its attempt to improve its statistical features has created many other normal distribution functions, such as NORM.DIST and NORM.S.DIST. But pay attention that we want to use the NORMSDIST function only with the Z-score. I agree that perhaps using these new formulas would make the model more 'efficient' (in this case, with less intermediate calculations, such as the Z-score). However, the reason I prefer this modelling technique is that I want you to understand what the normal curve is, what the standard deviation means, and how to go about standardising construction curves.

DRAWS

This is the actual percentage of the total construction cost incurred for the month.

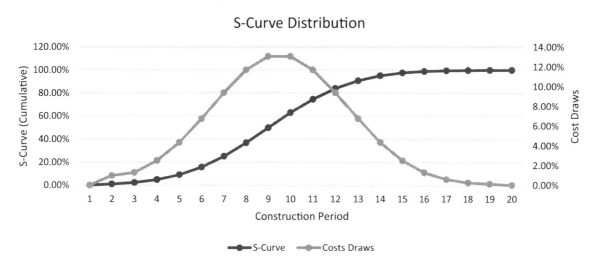

Figure 8.17 S-Curve Distribution.

A	B	C	E	F	G	H	I	J
91	**Hard Cost Distribution**	Check						
92	z-score			0.00	0.00	0.00	-2.25	-1.50
93	s-curve			0.00%	0.00%	0.00%	0.00%	1.22%
94	draw	0.0%		0.00%	0.00%	0.00%	0.00%	1.22%

Figure 8.18 Hard Cost Table 1.

A	B	C	DE	F
91	**Hard Cost Distribution**	Check		
92	z-score		= IF(F86=const_status,	(F87-const_dur/2)/(const_dur/6), 0)
93	s-curve		=IF(F84<=const_start,	0, IF(F84>=const_end, 1, NORMSDIST(E92)))
94	draw	=1-SUM(F94:Z94)	=F93-E93	

Figure 8.19 Hard Cost Table 2.

Known-Curve

The 'known' curve is one that has been estimated from the contract or through a quantity surveying firm, and as such, it is all hardcoded. You just need to enter the percentages per period and make sure they add up to 100%. Because there is no modelling required for calculating the 'known' curve, this cost curve will not be used in our models.

Pro-Forma Development Cash Flow

Now that we have calculated the cost distributions, we can then start modelling the pro-forma cash flow.

	B	C	E	F	G	H	I	J
97	PRO FORMA DEVELOPMENT CASH FLOW							
99	Period			0	1	2	3	4
100	Date			31-Dec-22	31-Mar-23	30-Jun-23	30-Sep-23	31-Dec-23
102	**Development Cash Flow**	*Check*						
103	Site Acquisition Costs	-		(1,064,000)	-	-	-	-
104	Hard Costs	-		-	-	-	-	(25,977)
105	Soft Costs	-		-	(181,667)	(181,667)	(181,667)	-
106	**Total Costs**			(1,064,000)	(181,667)	(181,667)	(181,667)	(25,977)
108	Office Sale			-	-	-	-	-
109	**Total Sales Proceeds**	-		-	-	-	-	-
111	**Total Development Cash Flow**	-		(1,064,000)	(181,667)	(181,667)	(181,667)	(25,977)
113	Total Funds Required	-		1,064,000	181,667	181,667	181,667	25,977

Figure 8.20 Pro Forma Development Cash Flow 1.

	B	C	DE	F
97	PRO FORMA DEVELOPMENT CASH FLOW			
99	Period		=F83	
100	Date		=F84	
102	**Development Cash Flow**	*Check*		
103	Site Acquisition Costs	=total_site+SUM(F103:Z103)	=-(F99=0)*total_site	
104	Hard Costs	=hard_costs+SUM(F104:Z104)	=-hard_costs*F94	
105	Soft Costs	=soft_costs+SUM(F105:Z105)	=-soft_costs*F89	
106	**Total Costs**		=SUM(F103:F105)	
108	Office Sale		=IF(F$100=exit_date,sales_proceeds,0)	
109	**Total Sales Proceeds**	=sales_proceeds-SUM(F109:Z109)	=SUM(F108:F108)	
111	**Total Development Cash Flow**	=O7-SUM(F111:Z111)	**=F106+F109**	
113	Total Funds Required	=H76-SUM(F113:Z113)	=-MIN(F111,0)	

Figure 8.21 Pro Forma Development Cash Flow 2.

Net Present Value (NPV) of Cash Flows

Now that we have calculated the development cash flow, we can discount these cash flows to find the present value (today's value) of future cash flow amounts. The reason for doing so is because one dollar (or any monetary figure) in the future is worth less today given the impact of inflation, but most importantly, the impact of opportunity costs.

To discount a cash flow, you will use the formula:

NPV = Future Cash Flow/(1 + discount rate)^period

Assuming a discount rate of 15%, our cash flow would be:

	A	B	C	D	E	F	G	H	I
97		PRO FORMA DEVELOPMENT CASH FLOW							
99		Period				0	1	2	3
100		Date				31-Dec-22	31-Mar-23	30-Jun-23	30-Sep-23
110									
111		Total Development Cash Flow	–			(1,064,000)	(181,667)	(181,667)	(181,667)
114									
115		PV of Cash Flow				(1,064,000)	(175,429)	(169,405)	(163,588)

Figure 8.22 NPV 1.

	A	B	C	D	E	F	G
97		PRO FORMA DEVELOPMENT CASH FLOW					
99		Period				=F83	=G83
100		Date				=F84	=G84
110							
111		Total Development Cash Flow	=O7-SUM(F111:Z111)			=F106+F109	=G106+G109
114							
115		PV of Cash Flow				=F111*(1+$O$13)^-(F99/4)	=G111*(1+$O$13)^-(G99/4)

Figure 8.23 NPV 2.

Ungeared Development Returns

The ungeared development returns will show you the returns assuming 100% equity, without any debt finance and are:

Profit: Sales Proceeds – Total Costs
Profit on cost: Profit/Total Costs
Equity: Because we have no debt finance in the ungeared calculations, it is assumed 100% equity. So, in this case, Equity = Total Costs (excluding debt finance costs).
Equity multiple: Profit/Total Costs + 1 (for each monetary amount you put in, how much will you take out).
Target POC: The benchmark for the profit on cost (POC). For example, the developer may be targeting a POC of 50%; in this case, the development would only be viable if the POC achieved is greater than 50%.
Residual on POC: Using goal seek or solver, what would be the maximum amount you could pay for the site to achieve your profit on cost (POC) benchmark. This maximum amount is also referred to as the residual value.

Target rate: The discount rate, hurdle rate of target IRR for the project. In other words, what should be the expected annual return for the project given its risk profile. In projects with duration greater than one year, this should be the preferred method for analysing viability.

IRR: The internal rate of return for this project given the timeliness of the cash flows, from planning to construction to exit. This can be done using the Excel functions =IRR or =XIRR.

NPV: This is the sum of the present values of the development cash flows. This can be calculated explicitly by discounting each cash flow and adding them up or using the Excel functions =NPV or =XNPV.

Residual on IRR: Using Goal Seek or Solver, what would be the maximum amount you could pay for the site in order to achieve your target rate benchmark.

	Un-Geared
Profit	2,197,626
Profit on Cost	58.85%
Equity	3,734,000
Equity Multiple	1.59
Target POC	50.0%
Res. on POC	1,207,160
Target Rate	15.0%
IRR	15.7%
NPV	65,594
Res. on IRR	1,061,648

Figure 8.24 Ungeared Development Returns 1.

	Un-Geared
Profit	=H31-H76
Profit on Cost	=O7/H76
Equity	=H76
Equity Multiple	=O7/O9+1
Target POC	0.5
Res. on POC	1207159.55187227
Target Rate	0.15
IRR	=(1+IRR(F111:Z111))^4-1
NPV	=SUM(F115:Z115)
Res. on IRR	1061648.25081561

Figure 8.25 Ungeared Development Returns 2.

To calculate the residual land value using a target profit on cost, you will need to use either the Goal Seek function or the SOLVER function. You will need to re-run the Goal Seek/Solver every time you change assumptions. So, I would leave it as a last step when setting up the development model.

Development Valuation and Analysis 179

Residual Value on Profit on Cost

GOAL SEEK:

Data > What If Analysis > Goal Seek:

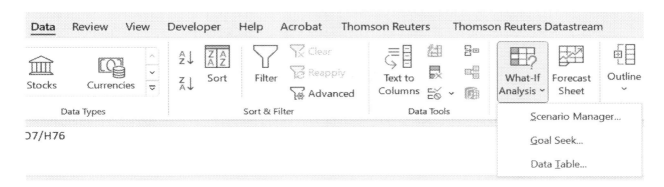

Figure 8.26 What-If Analysis 1.

Then, **Set Cell** Profit on Cost (**O8**) **To Value** 50% **By changing cell** site value (H36) as below:

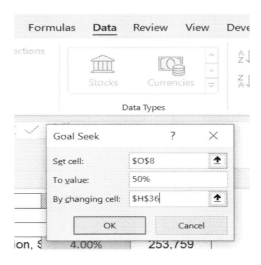

Figure 8.27 What-If Analysis 2.

The site value will change in the source cell, i.e., H36 and not appear in cell O12:

Figure 8.28 What-If Analysis 3.

You will then need to copy and paste the value calculated by Goal Seek into cell O12.

SOLVER:

To use Solver, follow the below steps:

1 Find the **Solver** function as in **Data** tab, then **Solver** (on the right-hand corner of the ribbon).

Figure 8.29 Solver 1.

2 If you cannot see SOLVER, go to FILE.

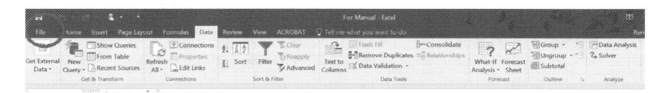

Figure 8.30 Solver 2.

3 Then, **Options** 4. Then, **Add-ins**.

Figure 8.31 Solver 3.

5 Then, select **Manage** > **Excel Add-ins** from the dropdown menu.

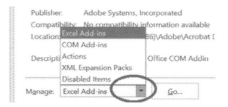

Figure 8.32 Solver 4.

6 Click **Go**.
7 Tick the **Solver Add-in** box and then click **OK**.

Figure 8.33 Solver 5.

Now that Solver has been inserted into your Excel, you can go to Step 1 and find the site value by targeting profit on cost:

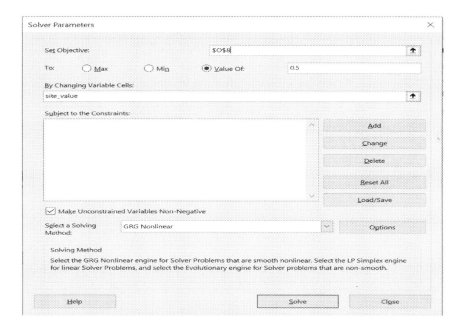

Figure 8.34 Solver 6.

As with Goal Seek, you will need to copy and paste the site value from H36 to O12.

Residual Value on IRR

The process is the same as to find the residual site/land value on profit on costs.

GOAL SEEK:

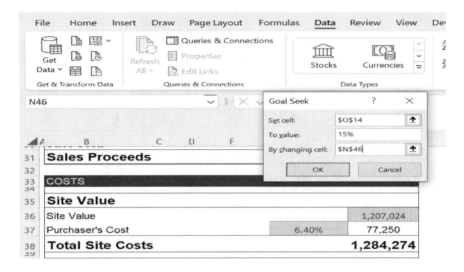

Figure 8.35 Residual Value on IRR.

SOLVER:

The Solver function to find the land value by targeting IRR will be:

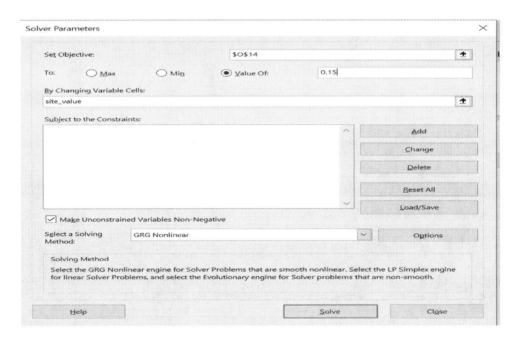

Figure 8.36 Residual Value on IRR 2.

Goal Seek vs Solver

They both fill the same function of trying to find an input to give a certain output. In our case, trying to find the site value to achieve a target profit on cost of internal rate of return.

However, Solver is more accurate as it runs for iterations to find the exact outcome. For example, if you run Goal Seek to find the site value based on a target IRR, the site value may give a resulting NPV different to zero, when we expect the NPV to be zero if the IRR is set to be the discount rate. If you run Solver for the same iteration, the result of NPV will be closer to zero than the result from Goal Seek. So, Solver is better! Moreover, Solver will 'remember' your last Solver iteration, so if you are to change assumptions many times and run the residual value, in Solver, you won't need to re-enter the inputs, just click on Solver and re-run it without the need to change any of the Solver inputs.

The only problem with Solver is that if you want to write macros, it doesn't work as well as with Goal Seek; so, my 'golden rule' is to run Solver for everything with the exception of when writing macros for VBA.

Conclusion

This chapter taught you how to create a real estate development appraisal using both the 'Profit and Loss' method, targeting the profit on cost, and the 'DCF' method, targeting the IRR. It goes without saying that the DCF method should always be preferred, because it considers the time value of money.

We have calculated the gross development value (GDV) and further deducted costs such as the hard and soft costs in order to derive a residual land value (RLV) using Goal Seek and SOLVER.

In this chapter, we analysed the development without gearing; as such, we have not considered any finance cost, either nominal or based on the DCF. In the next chapter, we will add debt (both senior and mezz/junior) and outside equity providers forming a joint-venture and splitting the profit according to hurdle rates based on IRR.

Chapter 9

Geared Development Investment

Chapter Contents

Introduction	184
Commercial Real Estate Debt and Development Activity	184
Senior Debt Finance	185
Cash Flow Waterfall – Equity First	185
Total Development Costs and Sources of Funds	186
Development Funding Financial Modelling	186
Equity and Debt Draws	186
Development Debt Structures	187
Mezzanine Finance	189
Mezzanine Debt Modelling	189
Total Debt Finance Cost	190
Preferred Returns Cash Flow	192
Geared Development Returns and JV Partners' Returns	193
Geared vs Ungeared Development Returns	194
Profit	194
Are Geared Equity Multiples and IRR Higher than the Ungeared Ones?	194
Conclusion	194

Introduction

Commercial Real Estate Debt and Development Activity

Development projects require considerable capital expenditure, with costs occurring at the beginning of the timeline and revenues at the end of the lifecycle of the development project. The capital intensity and timing of cash flows associated with development project means most developments would not be viable without short-term finance to support the land acquisition, planning, construction, and delivery of the property.

Development activity is inherently riskier than the ownership of income-producing, stabilised assets and tends to have a greater dependence on debt finance. Heightened risk aversion and credit drought for commercial real estate are most acute for the development sector and given its high socio-economic impact, its scarcity is deeply negative for the economy.

Development lending is usually expressed as a proportion of total costs, referred to as LTV (loan to cost) and will vary with the strength and track record of the borrower and the degree of risk mitigation achieved such as pre-letting achieved or forward sale of the completed project.

Large real estate companies with strong balance sheets benefit from the opportunity to raise debt capital through corporate banking facilities and/or the corporate bond markets. For those developers that require conventional development finance loans, the cost of funding will be higher even for those organisations with the strongest credit ratings. The cost to smaller and medium enterprises will be higher still, reflecting the greater risk associated with less substantial sponsors compared to large listed and private companies. The provision of debt is in most cases a prerequisite to development activity, and in turn, real estate development is pivotal to economic development.

The typical capital structure of development projects and the one that we will adopt in this course will be as in the figure below.

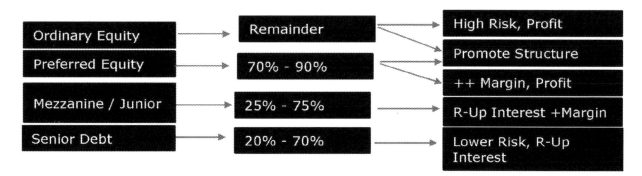

Figure 9.1 Typical Capital Structure of Development Projects.

Senior Debt Finance

In development debt finance, the type of senior debt used is mainly to acquire the site or land, as it has first charge. The typical structure of a senior loan is that of rolled up interest, given that the asset cash flow will be negative during development and will only become positive when the construction period is finalised or even after that if no pre-letting or forward sales are agreed prior to PC (practical completion).

From the senior debt section on rolled up interest, you know how this works; the difference now being that the drawdowns are more frequent and not a one-off as it is for the assets already built and in operation.

Cash Flow Waterfall – Equity First

The idea of equity first cash flow waterfall is that equity will be used up first, then senior debt, then mezzanine debt. The order may change, for example, equity first, mezzanine, and then senior. But our model will be able to cope with any orders; we will only use equity > senior > mezzanine as order of drawdowns because this is what we have seen most frequently in the debt market, since senior debt has first charge and will typically be used to buy the land. Mezzanine finance can also be structured in the form of unsecured, bridge-loan type of credit and will be drawn down once the site has been purchased and will be used to cover construction costs.

Illustratively, our development costs and funding will follow the chart on total development costs and sources of funds.

Total Development Costs and Sources of Funds

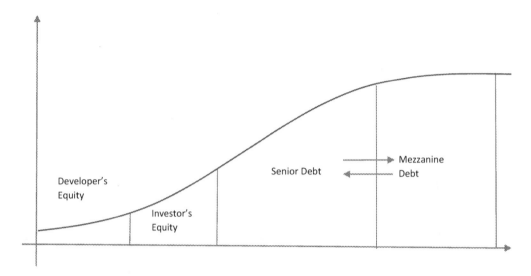

Figure 9.2 Total Development Costs and Sources of Funds.

Development Funding Financial Modelling

We will now start a new sheet called 'Dev Debt' which will then add third-party finance such as debt finance and equity investors.

The sources and costs of funds are:

Sources of Funds	Percent of Total Cost	Interest Rate	Origination Fee	Rolled Fee
Developer Equity	3.0%			
Investor Equity	17.0%			
Senior Debt	50.0%	6.0%	0.5%	Yes
Mezzanine Debt	30.0%	12.0%	2.0%	Yes
Total Sources	100.0%			

Figure 9.3 Sources and Costs of Funds.

Equity and Debt Draws

The total funds required (development costs) have been calculated, as well as the sales price. Noteworthy is that funds required are exclusive of financing fees, such as origination fees. The reason why I am telling you this is noteworthy is because although the percentage of senior debt is 50%, senior debt LTV may be different because the debt amount will include the fee amount if fee is rolled up into the debt. The same rationale applies to the mezz debt.

To replicate the graph above in a model, we will create formulas for the equity and debt draws as follows:

The lower of fund required minus any other funds used AND the available facility amount minus How much of the facility was already used?

Geared Development Investment

In Excel language:

= MIN (Fund Required - SUM(Previous Facilities), MAX Draw − SUM(Used Amount))

	I	J	K	L	M	N
1						
2		**DEBT AND EQUITY SCHEDULES - DEVELOPMENT FINANCE**				
3						
4						
5		Period				=Development!F83
6		Date				=Development!F84
7						
8		Total Funds Required	=SUM(N8:A)			=Development!F113
9		Sales Revenue				=Development!F109
10						
11		Equity & Debt Draws:	Max Draw:			
12		=B6	=D6*$L$8			=MIN(N$8-SUM(N$11:N11), $L12-SUM($M12:M12))
13		=B7	=D7*$L$8			=MIN(N$8-SUM(N$11:N12), $L13-SUM($M13:M13))
14		=B8	=D8*$L$8			=MIN(N$8-SUM(N$11:N13), $L14-SUM($M14:M14))
15		=B9	=D9*$L$8			=MIN(N$8-SUM(N$11:N14), $L15-SUM($M15:M15))
16		Total Capital Drawn:	=SUM(L12:L			**=SUM(N12:N15)**

Figure 9.4 Sources and Costs of Funds in Model.

Development Debt Structures:

Now that we know the amount that each fund source needs to contribute and when to the development project, we can create the debt schedules.

Senior Debt

As we know the amount that we will need to draw down from each source of funds (senior debt, mezzanine, equity), we can now go ahead and create the debt schedules with their drawdowns, interest calculation and repayments. If you need further explanation on the elements of the debt schedule, please refer to the Section 2, Chapter 6 of this book.

	I	J	K	L	M	N	O	P	Q
19									
20		**Senior Debt**							
21		Debt Brought Forward				-	503,305	692,521	884,576
22		Draw				493,419	181,667	181,667	181,667
23		Interest Due				-	7,550	10,388	13,269
24		Interest Paid				-	-	-	-
25		Fees Due				9,886	-	-	-
26		Fees Paid				-	-	-	-
27		Debt Repayment				-	-	-	-
28		Debt Carried Forward		Cost of Debt		503,305	692,521	884,576	1,079,511
29		**Debt Cash Flow**		(453,335)		493,419	181,667	181,667	181,667

Figure 9.5 Senior Debt 1.

	I	J	K	L	M	N
19						
20		**Senior Debt**				
21		Debt Brought Forward				=M28
22		Draw				=N$14
23		Interest Due				=N21*senior_rate/4
24		Interest Paid				=-IF(N$6=exit_date,SUM($N23:N23),0)
25		Fees Due				=IF(N$5=0,$L$14*senior_fee,0)
26		Fees Paid				=-IF(N$6=exit_date,SUM($N25:N25),0)
27		Debt Repayment				=-IF(N$6=exit_date,SUM(N21:N26),0)
28		Debt Carried Forward	Cost of Debt			=SUM(N21:N27)
29		**Debt Cash Flow**	=SUM(N29:AH29)			=N22+N24+N26+N27

Figure 9.6 Senior Debt 2.

Effective Senior Debt Cost

In order to calculate the effective senior debt cost, remembering that the interest rate is given on a nominal basis, and on a rolled-up basis, the effective rate should be calculated based on the IRR of the senior debt cash flow as this will consider both the effect of compounding and all fees that are added to the debt facility.

In our case, it is:

= IRR (debt cash flow)

Because our cash flow is quarterly, the annual rate will be:

= (1 + IRR(debt cash flow)) ^4 − 1

As such, from an equity perspective (the developer's perspective), the IRR of the debt cash flow is a cost, but from the lender's perspective, this will be their actual return on the facility.

In our model, the EAR (effective annual rate) for senior debt will be:

= (1 + IRR(N29:AH29))^4-1

The reason why I do not use the XIRR in this case is because sometimes senior debt may not be drawn down at period zero (0) in which case, the XIRR gives an error.

In a similar fashion, in order to calculate the cash amount (as opposed to an IRR figure), in pounds but it can also be in all you need to do is to add up the senior debt cash flow.

Total Cost of Senior Debt = SUM (Debt Cash Flow)

In our case:

Cost of Senior = SUM (N29:AH29)

You may find it similar to calculating profit, and you are right because the total cost of debt for the developer is equivalent to a 'profit' calculation for the senior lender. The reason why I put 'profit' in quotation marks is because we are not considering the lender's own cost of funding, so we do not know, from this calculation, what their actual profitability margin may be.

Mezzanine Finance

Mezzanine finance is a loose term in that it refers to the intermediate finance between the senior and common or ordinary equity. As such, some practitioners refer to junior debt finance as 'mezz', but preferred equity can also be referred as 'mezz', so it is important to read the terms of the financial agreement before committing any capital as they may have different meanings from provider to provider.

In our case, we will use the term 'mezz debt' to refer to the junior debt tranche. It therefore has lower priority return over senior debt and is riskier. This additional tranche is subordinated (second charge) to the senior debt but has higher priority to equity. Mezz debt has a higher required interest rate than senior debt and may have additional return features, such as profit share or preferred returns.

Mezzanine Debt Modelling

The figures for mezzanine debt will illustrate both the resulting figures as well as the formulas used in the calculations. Similarly to the senior debt, this facility will be on a rolled-up interest basis.

	I	J	K	L	M	N	O	P	Q
32									
33		**Mezzanine**							
34		Mezz Brought Forward				-	23,726	24,437	25,171
35		Draw				-	-	-	-
36		Interest Due				-	712	733	755
37		Interest Paid				-	-	-	-
38		Fees Due				23,726	-	-	-
39		Fees Paid				-	-	-	-
40		Capital Repayment				-	-	-	-
41		Mezz Carried Forward				23,726	24,437	25,171	25,926
42		**Mezz Cash Flow**		(381,137)		-	-	-	-
43									
44		Cash Flow Available For Distribution				-	-	-	-

Figure 9.7 Mezzanine Debt 1.

	I	J	K	L	M	N
32						
33		**Mezzanine**				
34		Mezz Brought Forward				=M41
35		Draw				=N$15
36		Interest Due				=N34*mezz_rate/4
37		Interest Paid				=-IF(N$6=exit_date,SUM($N36:N36),0)
38		Fees Due				=IF(N$5=0,$L$15*mezz_fee,0)
39		Fees Paid				=-IF(N$6=exit_date,SUM($N38:N38),0)
40		Capital Repayment				=-IF(N$6=exit_date,SUM(N34:N39),0)
41		Mezz Carried Forward				=SUM(N34:N40)
42		**Mezz Cash Flow**		=SUM(N42:AH42)		=N35+N37+N39+N40
43						
44		Cash Flow Available				=IF(N6=exit_date, N9+N29+N42, 0)

Figure 9.8 Mezzanine Debt 2.

As you can see, the formulas are absolutely the same as for the senior debt schedule, and you will just need to 'copy and paste' the formulas and change the references to mezz debt inputs instead.

Total Debt Finance Cost

The total finance costs using the DCF of senior and mezzanine will be the sum of both cash flows:

= Senior Cash Flow + Mezz Cash Flow

If you do that, you can then derive the effective borrowing costs in money terms and IRR.

A parenthesis here goes to the total debt finance cost according to some valuation techniques. Mainly, the notional finance costs.

Notional Finance Costs

These are the 'valuation' finance costs in which the assumption is that of a 100% gearing, representing a (fictional) loan to cost (LTC) of 100%. Moreover, the debt facility is assumed to be split in two: one for land acquisition and one for the remaining development costs (hard and soft costs); and some of the time, the interest applied will be the net initial yield used to calculate the gross development value (GDV). Consequently, these costs are not a projection of the actual debt finance costs as we have calculated using the cash flows for senior + mezzanine, but purely illustrative.

My view is that this cost is used to calculate the opportunity cost when valuers use the Profit on Cost method to calculate the residual land value, as it is only in this fictional debt cost that timing will have an impact on land value.

The formulas for calculating the notional finance costs are:

Land Debt Costs = Land Costs × ((1+ Assumed Interest Rate /12)^(total development duration in months) – 1)

Development Debt Costs (excluding Land) = (Total Development Costs – Land Costs) × ((1+ Assumed Interest Rate/12) ^ **(total development duration in months/2) – 1)**

- Note that the development debt costs will have the compounding factor (the total development duration) divided by 2. The reason for this is to take into account that the drawdowns for the development costs do not happen at the beginning of development as a one-off but in tranches, saving the developer some costs in accrued interest; in this case, half of what would have been accrued if all the drawdown was assumed on day 1.

Joint-Venture – Promote Waterfall

According to a joint-venture agreement, profits can be distributed using a subordinated cash flow, i.e., waterfall. The distributions are often not pari passu, meaning that the percentage of capital provided in the investment will not be proportional. In our example, we will use the proportions below:

In the structure above, the developer will provide 15% of the equity and the investor will provide the remainder of 85%.

For the first 10% IRR, the investor will retain 85% and the sponsor 15%. From 10% to 15%, the developer will receive a 10% promote, and from 15%, the developer will receive a 20% 'top-up'.

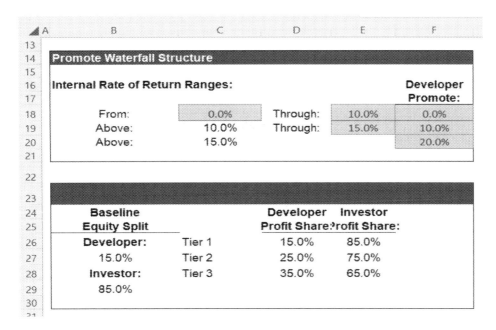

Figure 9.9 Joint Venture – Promote Waterfall.

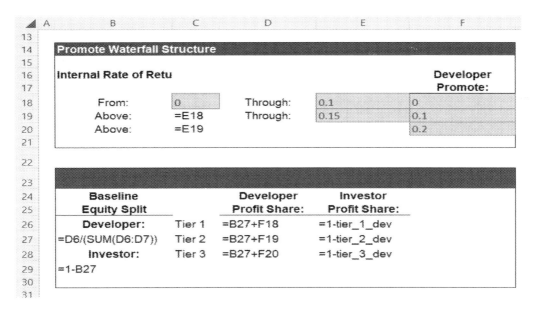

Figure 9.10 Joint Venture – Promote Waterfall 2.

In joint-venture agreements, the developer will be incentivised to obtain a very profitable project. The investor's preferences for low risk will be rewarded as the initial return will ensure the baseline.

Sometimes, the developer may even provide zero equity but a percentage of the upside above a certain return. This is what is also called 'sweat equity'. Also, the developer will typically receive a development, construction, and/or management fee.

Preferred Returns Cash Flow

1 Tiered Cash Flow

	I	J	K	L	M	N	O	P	Q	R
48		**Preferred Return**								
49		Tier 1 @ 10%								
50		Equity B/F				-	790,855	809,925	829,455	849,457
51		Equity Contribution			790,855	-	-	-	-	-
52		Accrual (FV)			-	-	19,070	19,530	20,001	20,484
53		Repayments			-	-	-	-	-	-
54		Equity C/F				790,855	809,925	829,455	849,457	869,940
55		Overflow		Check		-	-	-	-	-
56		Tier 1 Cash Flow		10.00%		790,855	-	-	-	-
57										
58		Tier 2 @ 15%								
59		Equity B/F				-	790,855	818,976	848,097	878,254
60		Equity Contribution			790,855	-	-	-	-	-
61		Accrual (FV)			-	-	28,121	29,121	30,157	31,229
62		Repayments			-	-	-	-	-	-
63		Equity C/F				790,855	818,976	848,097	878,254	909,483
64		Overflow		Check		-	-	-	-	-
65		Tier 2 Cash Flow		15.00%		790,855	-	-	-	-

Figure 9.11 Tiered Cash Flow.

	I	J	K	L	M	N
48		**Preferred Return**				
49		="Tier 1 @ " &tier_1*100&"%"				
50		Equity B/F				=M54
51		Equity Contribution				=N12+N13
52		Accrual (FV)				=N50* ((1+tier_1)^0.25-1)
53		Repayments				=-IF(N$6=exit_date,SUM(N50:N52),0)
54		Equity C/F				=SUM(N50:N53)
55		Overflow		Check		=N44+N53
56		Tier 1 Cash Flow		=(1+IRR(N56:AH56))^4-1		=N51+N53
57						
58		="Tier 2 @ " &tier_2*100&"%"				
59		Equity B/F				=M63
60		Equity Contribution				=N51
61		Accrual (FV)				=N59* ((1+tier_2)^0.25-1)
62		Repayments				=-IF(N$6=exit_date,SUM(N59:N61),0)
63		Equity C/F				=SUM(N59:N62)
64		Overflow		Check		=N44+N62
65		Tier 2 Cash Flow		=(1+IRR(N65:AH65))^4-1		=N60+N62

Figure 9.12 Tiered Cash Flow 2.

2 JV Partners' Cash Flows

	I	J	K	L	M	N	O	P	Q	R	S	T
67		**Cash Flow to Each Tier**										
68		Tier 1 Split				-	-	-	-	-	-	-
69		Developer				-	-	-	-	-	-	-
70		Investor				-	-	-	-	-	-	-
72		Tier 2 Split				-	-	-	-	-	-	-
73		Developer				-	-	-	-	-	-	-
74		Investor				-	-	-	-	-	-	-
76		Overflow				-	-	-	-	-	-	-
77		Developer				-	-	-	-	-	-	-
78		Investor				-	-	-	-	-	-	-
80		**Promote Cash Flow**										
81		*Developer*										
82		Equity Provided				(118,628)	-	-	-	-	-	-
83		Distributions				-	-	-	-	-	-	-
84		Cash Flow				(118,628)	-	-	-	-	-	-
86		*Investor*										
87		Equity Provided				(672,227)	-	-	-	-	-	-
88		Distributions				-	-	-	-	-	-	-
89		Cash Flow				(672,227)	-	-	-	-	-	-
92		**Equity Cash Flow**				(790,855)	-	-	-	-	-	-

Figure 9.13 JV Partners' Cash Flows 1.

Geared Development Investment

	I	J	K	L	M	N	O
67		**Cash Flow to Each T**					
68		Tier 1 Split				=-N53	=-O53
69		Developer				=N68*tier_1_dev	=O68*tier_1_dev
70		Investor				=N68*tier_1_investor	=O68*tier_1_investor
72		Tier 2 Split				=-N62-N68	=-O62-O68
73		Developer				=N72*tier_2_dev	=O72*tier_2_dev
74		Investor				=N72*tier_2_investor	=O72*tier_2_investor
76		Overflow				=N64	=O64
77		Developer				=N76*tier_3_dev	=O76*tier_3_dev
78		Investor				=N76*tier_3_investor	=O76*tier_3_investor
80		**Promote Cash Flow**					
81		*Developer*					
82		Equity Provided				=-N12	=-O12
83		Distributions				=N69+N73+N77	=O69+O73+O77
84		Cash Flow				=SUM(N82:N83)	=SUM(O82:O83)
86		*Investor*					
87		Equity Provided				=-N13	=-O13
88		Distributions				=N70+N74+N78	=O70+O74+O78
89		Cash Flow				=SUM(N87:N88)	=SUM(O87:O88)
92		**Equity Cash Flow**				**=N84+N89**	**=O84+O89**

Figure 9.14 JV Partners' Cash Flows 2.

Geared Development Returns and JV Partners' Returns

Now that we have calculated the debt and equity providers' cash flows, we calculate their returns on the development sheet:

		N	O	P	Q	R
4	**RETURNS**					
6			Un-Geared	Geared	Developer	Investor
7		Profit	2,197,626	1,428,921	402,246	1,026,674
8		Profit on Cost	58.85%	31.7%		
9		Equity	3,734,000	746,800	112,020	634,780
10		Equity Multiple	1.59	2.91	4.59	2.62
11		Target POC	50.0%	30.0%		
12		Res. on POC	1,207,160	1,043,486		
13		Target Rate	15.0%	25.0%	30.0%	25.0%
14		IRR	15.7%	28.6%	43.1%	25.4%
15		NPV	65,594	96,150	56,638	8,925
16		Res. on IRR	1,061,648	1,144,443		

Figure 9.15 Geared Developments and JV Partners' Returns 1.

	M	N	O	P	Q	R
4		**RETURNS**				
6			Un-Geared	Geared	Developer	Investor
7		Profit	=H31-H76	=SUM('Dev Debt'!N92:AH92)	=SUM(dev_cf)	=SUM(inv_cf)
8		Profit on Cost	=O7/H76	=P7/H78		
9		Equity	=H76	=SUM('Dev Debt'!L12:L13)	=-SUMIF(dev_cf,"<0")	=-SUMIF(inv_cf,"<0")
10		Equity Multiple	=O7/O9+1	=P7/P9+1	=Q7/Q9+1	=R7/R9+1
11		Target POC	0.5	0.3		
12		Res. on POC	1207159.55187227	1043486.30443584		
13		Target Rate	0.15	0.25	0.3	0.25
14		IRR	=(1+IRR(F111:Z111))^4-1	=XIRR(equity_cf,cf_dates)	=XIRR(dev_cf,cf_dates)	=XIRR(inv_cf,cf_dates)
15		NPV	=SUM(F115:Z115)	=XNPV(P13,equity_cf,cf_dates)	=XNPV(Q13,dev_cf,cf_dates)	=XNPV(R13,inv_cf,cf_dates)
16		Res. on IRR	1061648.25081561	1144442.5278756		

Figure 9.16 Geared Developments and JV Partners' Returns 2.

Geared vs Ungeared Development Returns

Profit

The geared profit includes the cost of debt finance so it will inevitably be LOWER than the un-geared profit. There are several ways to calculate the geared profit, so it would be a good 'check' to calculate them all and see if they match.

> Geared Profit = Ungeared Profit − Cost of Debt
> Geared Profit = SUM of Equity Cash Flow
> Geared Profit = Sales Proceeds − Total Costs (including debt finance)

Note that the geared profit on cost is also LOWER – that's why valuers use the assumption of 100% debt finance for residual valuations.

Are Geared Equity Multiples and IRR Higher than the Ungeared Ones?

NOT ALWAYS. If, for example, all your ungeared profit is used to pay back the interest and fees of your debt, then the return on equity will be lower, and consequently, the equity multiple will be lower as well. In terms of IRR, the geared IRR will only be greater than the ungeared IRR if the effective borrowing rate (including interest and fees) is lower than the ungeared IRR.

> Geared IRR > Ungeared IRR *if* Ungeared IRR > Effective Borrowing Rate

Conclusion

In this chapter, we have explained and modelled development debt and equity structures following an equity-first drawdown schedule and calculated equity and profit distribution accordingly to promote structures and preferred returns agreements, which are normally seen in joint-venture or partnership structures.

An important topic covered by this chapter is the analysis of geared development investment, showing how gearing can impact the returns and add risks. Knowing this helps developers make smarter choices about their capital structures.

In this chapter, we also introduced the concept of mezzanine finance as junior debt, which fills the gap between senior debt and equity, and we finally explained the modelling of joint-venture – promote waterfall structures, modelling profit distributions according to hurdle rates which are based on the financial performance of the project. Finally, we wrapped-up with the geared vs un-geared development returns, so you can analyse the impact of debt and different capital structures in real estate development investments.

Section 4

Risk Modelling

Measuring the riskiness of an investment is not always easy as some risks cannot be mathematically quantifiable, for example, reputation risk, but we will learn that there are different ways of showing risk.

In this section, I will show you how to produce risk-modelling techniques, mainly the sensitivity analysis, scenario analysis and the Monte Carlo simulation. I will use use the technique for Monte Carlo to do analyse climate risk and its impact on asset value and returns.

Chapter 10

Sensitivity Analysis

Chapter Contents

1-Way Data Table	197
Finding the IRRs and NPVs by Varying LTC from 30% to 80%, in 10% Steps	199
Graphs	200
Two-Way Data Table	202
Question: What Happens When the Capital Value and LTC Change at the Same Time?	203
Conditional Formatting	204
Scenario Analysis	208
Stress Testing	213
Setting Capital Requirements	213
Monte Carlo Simulation	213
Randomising the Variables	214
Automating the Returns Calculations	216
Analysing the Result	218
Histogram	220
Climate Risk Modelling	222
Climate Insurance	224
Climate Excess Costs	225
Climate Resilience CAPEX	225
Conclusion	226
Final Words	227

Sensitivity analysis is the process of checking how 'sensitive' your investment is to a certain variable. For example, if you change your exit yield from 5 to 6%, and your IRR changes from 20 to 10%, you will realise that your investment has a 1 to 10% sensitiveness.

As such, Sensitivity analysis involves changing one or more key assumptions to show how our outputs can change by varying inputs. Inputs that are typically examined in this type of analysis include the expected exit yield, market rental rate, vacancy rates, operating expenses, and expected rental growth.

Using the Data Table function in Excel, we can change one (1-way data table) or two inputs (two-way data table) at a time.

But if we are clever (and we are), we can create scenarios by changing multiple data points at the same time. So here, I will show you how to do the sensitivity analysis by changing one data point (1-way table) and two data points (two-way table). Then, I will show you how to change multiple inputs at the same time as a dynamic scenario analysis.

1-Way Data Table

Finding the IRRs and NPVs when capital values change from £250 to £450 per sqf in £25 steps. Steps:

1 Create a table as below:

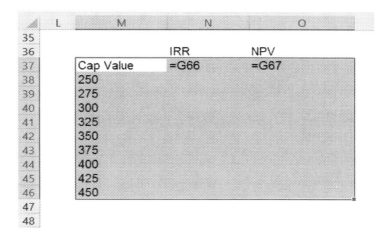

Figure 10.1 One-Way Data Table 1.

2 Highlight cells

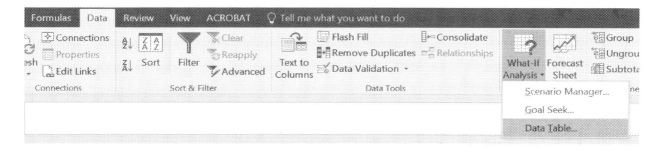

Figure 10.2 One-Way Data Table 2.

3 Go to DATA > WHAT-IF ANALYSIS > DATA TABLE as below:

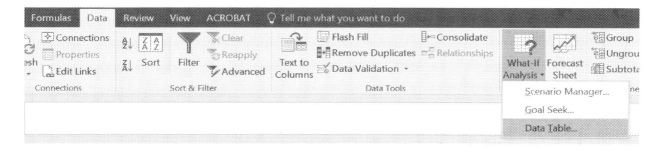

Figure 10.3 One-Way Data Table 3.

4 The DATA TABLE window will display:

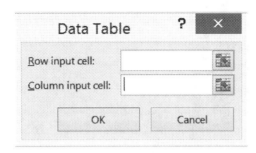

Figure 10.4 One-Way Data Table 4.

5 As we are adding values in a column, click in the COLUMN INPUT CELL field.
6 You will need to find in your sheet where the data point (capital value) that you want to change lies.

	B	C	D	E	F	G	H	I
1								
2	DEVELOPMENT APPRAISAL							
3								
4	REVENUES						PERIOD	
5	Sales Valuation	Units	ft2	Rate ft2	Unit Price	Gross Sales	Start	End
6	1 Beds	50	40,000	350.00	280,000	14,000,000	14	19
7	2 Beds	25	28,000	350.00	392,000	9,800,000	14	21
8	Total	75	68,000			23,800,000		

Figure 10.5 One-Way Data Table 5.

7 Link E7 to E6, so that both rates change their values in the sensitivity calculation, i.e.:
   Cell E7 = E6
8 Select the cell E6, and the DATA TABLE window will look like this:

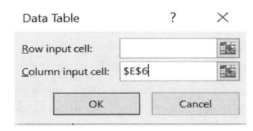

Figure 10.6 One-Way Data Table 6.

9 Click OK.

Sensitivity Analysis

10 The sensitivity table will look like this:

	L	M	N	O
35				
36			IRR	NPV
37		Cap Value	274.06%	2,478,270
38		250	#NUM!	-1,283,781
39		275	#NUM!	-343,268
40		300	127%	597,244
41		325	206%	1,537,757
42		350	274%	2,478,270
43		375	337%	3,418,783
44		400	398%	4,359,296
45		425	457%	5,299,809
46		450	515%	6,240,321
47				

Figure 10.7 One-Way Data Table 7.

Note that you can add as many output variables as you want.

Note that for very low capital values such as £250 and £275, there is no solution for IRR. Therefore, you should rely on NPV.

Finding the IRRs and NPVs by Varying LTC from 30% to 80%, in 10% Steps

Steps:

1 Create a table as below:

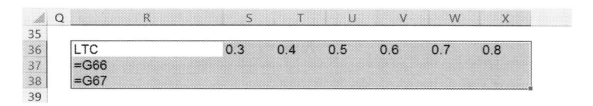

Figure 10.8 Finding IRR and NPV by Varying LTC.

2 Highlight cells.
3 Then:

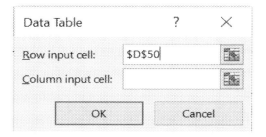

Figure 10.9 Finding IRR and NPV by Varying LTC 2.

4 The sensitivity table will be as below:

	R	S	T	U	V	W	X
35							
36	LTC	30%	40%	50%	60%	70%	80%
37	274.06%	64.61%	82.51%	112.15%	165.80%	274.06%	531.11%
38	2,478,270	618,281	1,083,278	1,548,276	2,013,273	2,478,270	2,943,267
39							

Figure 10.10 Finding IRR and NPV by Varying LTC 3.

Graphs

Creating graphs is a great way to visualise data and make sense of it. When creating sensitivity data, it is best practice to use graphs so you can see the rate of decline in IRR and NPV or in other financial measures. When modelling uncertainty, we are interested in both the rate and magnitude of the change.

In the example below, we will show how to create a column chart using the data table we have just created.

Steps:

1. Select the data you want to chart. In this case, it will be the NPVs:

	K	L	M	N	O
35					
36				IRR	NPV
37			Cap Value	274.06%	2,478,270
38			250	#NUM!	-1,283,781
39			275	#NUM!	-343,268
40			300	127%	597,244
41			325	206%	1,537,757
42			350	274%	2,478,270
43			375	337%	3,418,783
44			400	398%	4,359,296
45			425	457%	5,299,809
46			450	515%	6,240,321
47					

Figure 10.11 Graphs 1.

2 Press F11.
3 The following will appear:

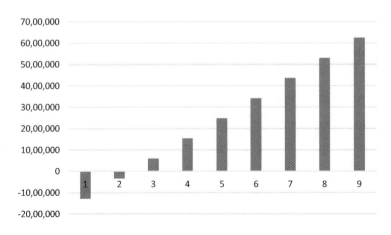

Figure 10.12 Graphs 2.

4 Right click, SELECT DATA.

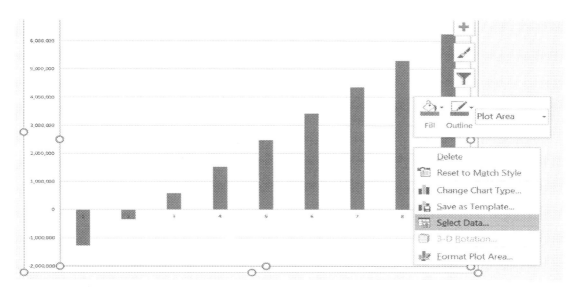

Figure 10.13 Graphs 3.

5 The following window will display:

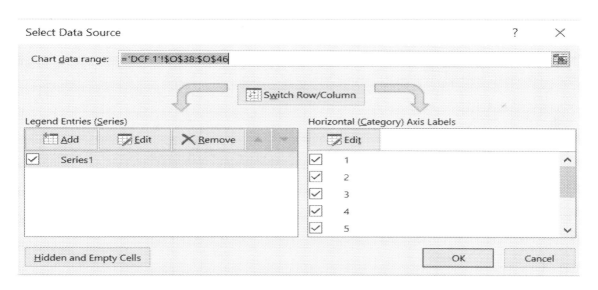

Figure 10.14 Graphs 4.

6 In the HORIZONTAL (CATEGORY) AXIS LABELS, click EDIT.
7 The following window will display:

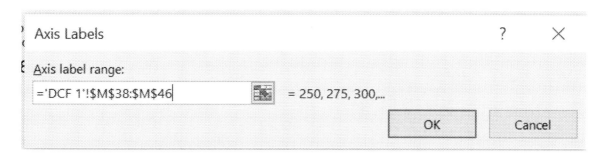

Figure 10.15 Graphs 5.

8 Select the Axis labels from your original sheet:
9 Click OK.
10 Add Chart Title and Axis Titles using the Quick Layout:
   DESIGN > QUICK LAYOUT

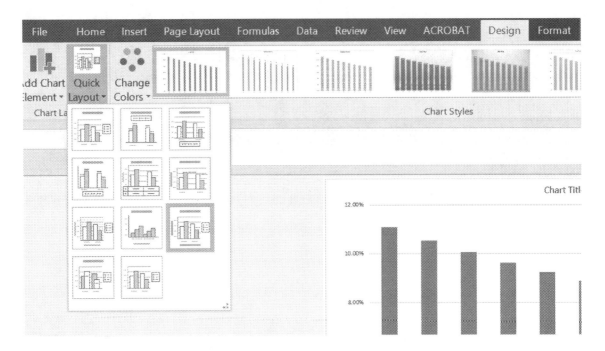

Figure 10.16 Graphs 6.

11 Select the Layout 9.
12 Edit text directly on the chart.

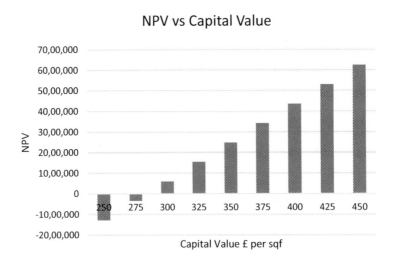

Figure 10.17 Graphs 7.

Two-Way Data Table

The two-way data tables work in the same way as one-way data tables except that you can vary two parameters at once.

With two-way data tables, you need to set up a column of data for one parameter and a row of data for the second parameter. The value to be changed (in our case, the IRR) needs to go at the intersection of the row and column.

Sensitivity Analysis 203

We will set up the data varying capital value and LTC.

Question: What Happens When the Capital Value and LTC Change at the Same Time?

Steps:

1 Set up the table as below, with LTC in a row and capital value in a column

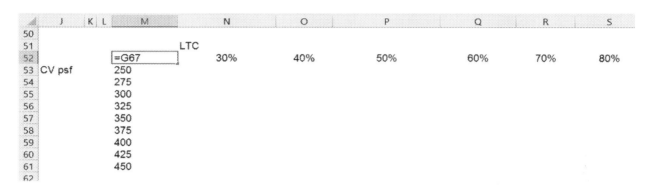

Figure 10.18 Two-Way Table 1.

2 In the intersection of both data series, select the variable you want to change. In our case, the NPV or cell M51.
3 Select the whole table.

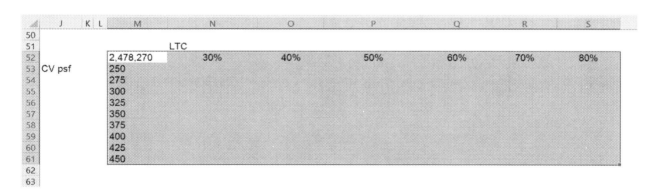

Figure 10.19 Two-Way Table 2.

4 Follow steps 3 and 4 as in the one-way data table example.
5 The DATA TABLE window will display:

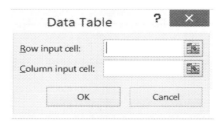

Figure 10.20 Two-Way Table 3.

204 Real Estate Financial Modelling in Excel

6 In the ROW INPUT CELL, enter D50 (LTC).
7 In the COLUMN INPUT CELL, enter E6 (capital value psf).

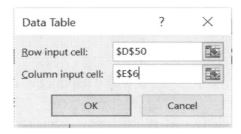

Figure 10.21 Two-Way Table 4.

8 Your two-way data table will look as follows:

			LTC					
		2,478,270	30%	40%	50%	60%	70%	80%
CV psf		250	-3,139,374	-2,675,475	-2,211,577	-1,747,679	-1,283,781	-819,883
		275	-2,199,960	-1,735,787	-1,271,614	-807,441	-343,268	120,905
		300	-1,260,546	-796,099	-331,651	132,797	597,244	1,061,692
		325	-321,133	143,590	608,312	1,073,035	1,537,757	2,002,480
		350	618,281	1,083,278	1,548,276	2,013,273	2,478,270	2,943,267
		375	1,557,695	2,022,967	2,488,239	2,953,511	3,418,783	3,884,055
		400	2,497,108	2,962,655	3,428,202	3,893,749	4,359,296	4,824,843
		425	3,436,522	3,902,343	4,368,165	4,833,987	5,299,809	5,765,630
		450	4,375,935	4,842,032	5,308,128	5,774,225	6,240,321	6,706,418

Figure 10.22 Two-Way Table 5.

Conditional Formatting

I will now show you how to do the conditional formatting for the table above. Changing the format of cells that have some specific meaning for your cash flow analysis is a helpful way to visualise risk.

Highlight all NPVs greater than 0.

Steps:

1 Select the data table output range, cells N53 to S61 as below:

			LTC					
		2,478,270	30%	40%	50%	60%	70%	80%
CV psf		250	-3,139,374	-2,675,475	-2,211,577	-1,747,679	-1,283,781	-819,883
		275	-2,199,960	-1,735,787	-1,271,614	-807,441	-343,268	120,905
		300	-1,260,546	-796,099	-331,651	132,797	597,244	1,061,692
		325	-321,133	143,590	608,312	1,073,035	1,537,757	2,002,480
		350	618,281	1,083,278	1,548,276	2,013,273	2,478,270	2,943,267
		375	1,557,695	2,022,967	2,488,239	2,953,511	3,418,783	3,884,055
		400	2,497,108	2,962,655	3,428,202	3,893,749	4,359,296	4,824,843
		425	3,436,522	3,902,343	4,368,165	4,833,987	5,299,809	5,765,630
		450	4,375,935	4,842,032	5,308,128	5,774,225	6,240,321	6,706,418

Figure 10.23 Conditional Formatting 1.

2 Then, go to HOME > CONDITIONAL FORMATTING > NEW RULE:

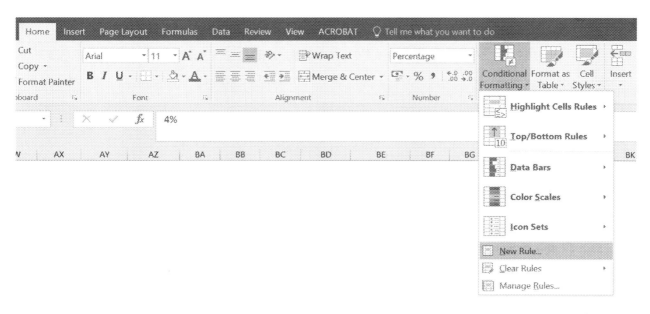

Figure 10.24 Conditional Formatting 2.

The NEW FORMATTING RULE window will display.

3 Select FORMAT ONLY CELLS THAT CONTAIN:

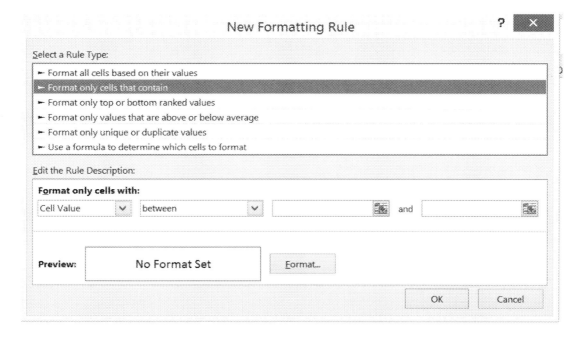

Figure 10.25 Conditional Formatting 3.

In the EDIT THE RULE DESCRIPTION, FORMAT ONLY CELLS WITH, change fields to CELL VALUE > GREATER THAN:

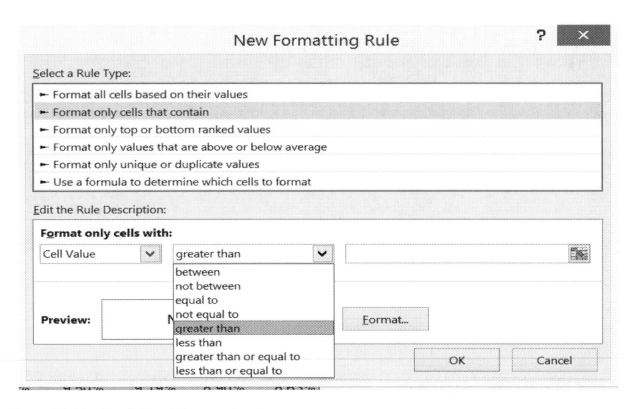

Figure 10.26 Conditional Formatting 4.

4 And add 0 in the field to the right (you want to find all scenarios where NPV > 0):

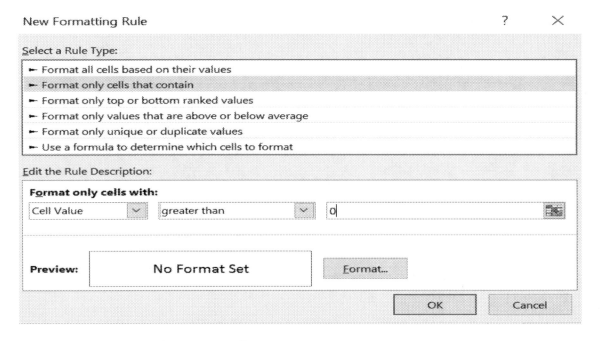

Figure 10.27 Conditional Formatting 5.

5 Click FORMAT.
6 The FORMAT CELLS window will appear. Click on the tab FILL.
7 Select the colour you wish to highlight cells with:

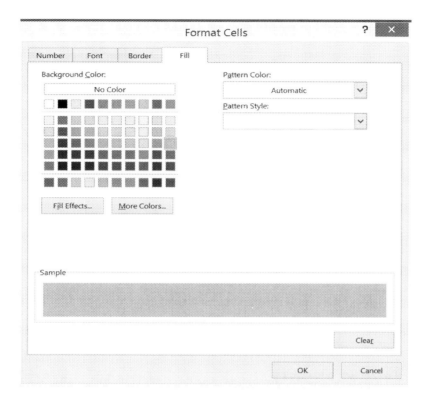

Figure 10.28 Conditional Formatting 6.

8 Your two-way data table will look like this:

	J	K	L	M	N	O	P	Q	R	S
50										
51					LTC					
52				2,478,270	30%	40%	50%	60%	70%	80%
53	CV psf			250	-3,139,374	-2,675,475	-2,211,577	-1,747,679	-1,283,781	-819,883
54				275	-2,199,960	-1,735,787	-1,271,614	-807,441	-343,268	120,905
55				300	-1,260,546	-796,099	-331,651	132,797	597,244	1,061,692
56				325	-321,133	143,590	608,312	1,073,035	1,537,757	2,002,480
57				350	618,281	1,083,278	1,548,276	2,013,273	2,478,270	2,943,267
58				375	1,557,695	2,022,967	2,488,239	2,953,511	3,418,783	3,884,055
59				400	2,497,108	2,962,655	3,428,202	3,893,749	4,359,296	4,824,843
60				425	3,436,522	3,902,343	4,368,165	4,833,987	5,299,809	5,765,630
61				450	4,375,935	4,842,032	5,308,128	5,774,225	6,240,321	6,706,418
62										

Figure 10.29 Conditional Formatting 7.

Notes for your sensitivity analysis:

- The sensitivity analysis assumes that the probability of occurrence of any results are the same. It means that the probability of you getting a 3.0% net yield in your office asset and 5 quarters of construction duration is the same as your base case. But in fact, you expect your base case to be the most probable scenario.
- When creating a two-way data table for sensitivity analysis, it doesn't tell you about the sensitivity of each variable alone on your returns. So, it may be best to stick with the one-way data table for you to check which variable(s) has(have) the most impact on your returns.
- Don't create a myriad of data tables, focus on a few variables that are relevant and that can provide you with meaningful insights into the risk profile of your investment. The more data tables you create, the heavier your model will become and the more you will lose sight on the most important variables to focus when doing the asset management.

Scenario Analysis

Scenario analysis is a way of looking at possible future outcomes by considering alternative possible events (sometimes called "alternative worlds"). Thus, scenario analysis is a method of producing projections under certain circumstances.

Let us consider the following scenarios for our projections:

	M	N	O	P	Q	R	S
28							
29		SCENARIO ANALYSIS					
30							
31		Scenario	Estimate	Office Yield	Retail Yield	Construction	
32		No.	Description	%	%	Period (qtr)	
33		1	Optimistic	-1.0%	-2.0%	-1	
34		2	Neutral	0.0%	0.0%	0	
35		3	Pessimistic	+1.0%	+2.0%	+1	

Figure 10.30 Scenario Analysis 1.

Question: What happens to the profitability projections in each scenario?

Steps:

1 Once you have set up your scenarios, you will need to link these variables to your INPUT tables.
2 I recommend that you use INDEX as this function is more flexible and robust when moving columns and rows around the model instead of VLOOKUP or HLOOKUP.
3 Linking the exit yield:
   On the cell of exit yield, you will use the INDEX function, which will then link the exit yield to the calculations depending on which scenario you are testing (the scenario number).

For instance, if you want to test Scenario 1 (optimistic), then the exit yield should be 5.00%, 2 (neutral) 6.00%, and 3 (pessimistic) 7.00%.

The syntax for the INDEX function is as follows:

=INDEX (lookup array, row, column)

In our case, this will be:

= INDEX (exit yield array, scenario no.)

You will need to do the same for rental growth and interest rate, i.e., replace the current inputs with an INDEX formula that will look up the values.

4 Now that the scenario variables have been linked to your model, you will need to add a 'switch' which will select the Scenario No. you want to test.
5 Note that the 'Active No' is a dropdown menu (from **Data Validation > List**).

Figure 10.31 Scenario Analysis 2.

The final result looks like this:

Figure 10.32 Scenario Analysis 3.

210   Real Estate Financial Modelling in Excel

6 Link the scenario entries with the input data. Note that you will create a new cell with the baseline input and the 'active' cell, i.e., the reference cell for the calculation of the cash flow will need to contain the scenario value. Therefore,

   For construction period:

	A	B	C	D	E	F	G	H	I
9		TIMESCALE							
10					Status No	Start Date	Duration	End Date	Base Line
11		Planning, Design, Pre Build			1	30-Jun-19	2 qtrs	31-Dec-19	
12		Construction			2	01-Jan-20	8 qtrs	31-Jan-22	=G12+R36
13		Lease Up			3	01-Feb-22	2 qtrs	31-Aug-22	
14		Stabilisation			4	01-Sep-22	2 qtrs	31-Mar-23	
15		Sale (Exit)			0	31-Mar-23			

Figure 10.33 Scenario Analysis 4.

For office and retail yields:

	A	B	C	D	E	F	G	H	I
16									
17		VALUATION							
18									
19		REVENUES			NIA	Rent	Net Yield	Gross	Base Line
20					ft2	ft2	%	Value	
21		Office Units			5,400	50.00	4.00%	6,750,000	=G21+P36
22		Retail Units			1,000	80.00	6.00%	1,333,333	=G22+Q36

Figure 10.34 Scenario Analysis 5.

6 Make sure that the 'Base Line' cells in **column I** are linked to your cash flows and NOT the input cells in **column G.**

   If you need to check this, when on cell **I12**, go to Formulas > Trace Dependents, and you should see the arrows going out of this cell.

	A	B	C	D	E	F	G	H	I
10					Status No	Start Date	Duration	End Date	Base Line
11		Planning, Design, Pre Build			1	30-Jun-19	2 qtrs	31-Dec-19	
12		Construction			2	01-Jan-20	8 qtrs	31-Jan-22	8 qtrs
13		Lease Up			3	01-Feb-22	2 qtrs	31-Aug-22	
14		Stabilisation			4	01-Sep-22	2 qtrs	31-Mar-23	
15		Sale (Exit)			0	31-Mar-23			

Figure 10.35 Scenario Analysis 6.

The same should happen with the yield cells.

7 To create the scenario projections, you will need to use the Excel function DATA TABLES (just like in the sensitivity analysis).

Where the standard Data Tables allow for varying 1 or 2 parameters, scenario analysis will need to use multiple variables. Thus, the process of creating scenarios can also be called a 'multi-way data table' technique.

The layout will be as follows:

	M	N	O	P	Q	R
4		**RETURNS**				
5						
6			Un-Geared			
7		Profit	3,006,305			
8		Profit on Cost	73.4%			
9		Target POC	30.0%			
10		Res. on POC	2,284,852			
11		Target Rate	15.0%			
12		IRR	23.6%			
13		NPV	729,224			
14		Res. on IRR	1,729,224			
28						
29		**SCENARIO ANALYSIS**				
30						
31		Scenario	Estimate	Office Yield	Retail Yield	Construction
32		No.	Description	%	%	Period (qtr)
33		1	Optimistic	-1.0%	-2.0%	-1
34		2	Neutral	0.0%	0.0%	0
35		3	Pessimistic	+1.0%	+2.0%	+1
36		2	Neutral	0.0%	0.0%	0
37						
38		Active No.	2			
39						
40			Description	IRR	NPV	Res. IRR
41		Scenario	=O36	=O12	=O13	=O14
42		1				
43		2				
44		3				

Figure 10.36 Scenario Analysis 7.

To capture the different outputs for each scenario, all that needs doing now is completing the DATA TABLE.

8 Highlight the table:

	M	N	O	P	Q	R
37						
38		Active No.	2			
39						
40			Description	IRR	NPV	Res. IRR
41		Scenario	=O36	=O12	=O13	=O14
42		1				
43		2				
44		3				

Figure 10.37 Scenario Analysis 8.

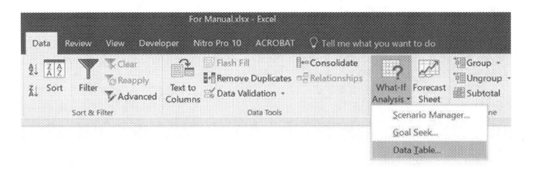

Figure 10.38 Scenario Analysis 9.

9 **Go to** Data > What-If Analysis > Data Table
10 Then, link the Data Table to the Scenario No. as below:

	Description	IRR	NPV	Res. IRR
Active No.	2			
Scenario	Neutral	23.57%	729,224	1,729,224
1				
2				
3				

Data Table
Row input cell:
Column input cell: $O$38

Figure 10.39 Scenario Analysis 10.

The scenarios will look like this:

SCENARIO ANALYSIS

Scenario No.	Estimate Description	Office Yield %	Retail Yield %	Construction Period (qtr)
1	Optimistic	-1.0%	-2.0%	-1
2	Neutral	0.0%	0.0%	0
3	Pessimistic	+1.0%	+2.0%	+1
2	Neutral	0.0%	0.0%	0

Active No.	2			
	Description	IRR	NPV	Res. IRR
Scenario	Neutral	23.57%	729,224	1,729,224
1	Optimistic	41.73%	2,425,316	3,425,316
2	Neutral	23.57%	729,224	1,729,224
3	Pessimistic	12.07%	-239,015	760,985

Figure 10.40 Scenario Analysis 11.

11 Now, look at the outputs and make inferences in terms of risk and return (the actual analysis). Questions to ask yourself:

What is the probability of each scenario occurring?
Are you better off weighting those outputs according to the probability of outcomes?
What are they telling you in terms of risk and return?
Can you plot them to make the interpretation easier?
Do you need to make many more assumptions for different scenarios?
Which variables should be added to the analysis? Are they important?

Stress Testing

Stressing models are just extensions to the scenario analysis models we have already addressed. There are mainly three forms of stressing models: **factor push models**, **maximum loss optimisation,** and **worst-case scenarios**.

In factor push analysis, you will push a factor (or factors) to the extreme and measure the impact on the project outputs. Maximum loss optimisation involves identifying risk factors that have the greatest potential for impacting the project outcomes. The worst-case scenario will push all risk factors to their worst cases to measure the absolute worst case for the project.

Setting Capital Requirements

To determine the allocation of capital to the project, you must determine the allocation in a way that maximises potential returns without replacing the viability of the project (bankruptcy). We know that capital set aside in form of cash or highly liquid risk-free assets, such as government bonds, will drag the returns downwards under 'normal' market conditions.

Some capital requirements are set by regulation (e.g., banks, pension, and insurance companies will have a minimum reserve requirement). Using stress testing, bank, pension, and insurance regulators set capital requirements such that the probability of insolvency is acceptable.

For the borrower, the cash reserve can be determined by the lender, and it is normally in months of debt service. This can range from 0 to 6 months depending on the profile of the investment, for instance whether the property is an income-producing asset or higher risk asset (development).

Monte Carlo Simulation

What if we could generate random numbers and see their impact on our returns?

What if these numbers were calculated 10, 100, infinite times, and you could investigate the risk associated with this random walk?

Can you imagine if we could collect all results and observe the behaviour of returns and compare the impact of randomness across investment products?

This is the idea of the Monte Carlo simulation, whereby you analyse the risk of the investment given a range of possible outcomes.

These outcomes depend on the shape of the probability distribution, and in our case, we will use the normal probability distribution that can be described by the mean and standard deviation of the distribution.

We will continue using the development investment example.

First, we need to save the workbook as a different file and make it 'macro enabled' (.xlsm).

Randomising the Variables

Then, we need to 'randomise' the construction quarter (or any quarter variables, such as rent-free, void periods) variables using the function:

= ROUND(NORMINV(RAND (),MEAN,STANDARD DEVIATION),0)

The reason you need the ROUND function is because in the model, quarters need to be exact numbers (integers). This type of data is called *discrete*.

	B	C	D	E	F	G	H	I	J
10				Status No	Start Date	Duration	End Date	Monte Carlo	Base Line
11	Planning, Design, Pre Bu			1	30-Jun-19	2 qtrs	31-Dec-19		
12	Construction			2	01-Jan-20	8 qtrs	31-Oct-21	=ROUND(NORMINV(RAND(),0,1),0)	7 qtrs
13	Lease Up			3	01-Nov-21	2 qtrs	31-May-22		
14	Stabilisation			4	01-Jun-22	2 qtrs	31-Dec-22		
15	Sale (Exit)			0	31-Dec-22				

Figure 10.41 Monte Carlo Simulation 1.

The baseline is therefore the construction duration plus the Monte Carlo.

	B	C	D	E	F	G	H	I	J
10				Status No	Start Date	Duration	End Date	Monte Carlo	Base Line
11	Planning, Design, Pre Bu			1	30-Jun-19	2 qtrs	31-Dec-19		
12	Construction			2	01-Jan-20	8 qtrs	31-Jan-22	0	=G12+I12
13	Lease Up			3	01-Feb-22	2 qtrs	31-Aug-22		
14	Stabilisation			4	01-Sep-22	2 qtrs	31-Mar-23		
15	Sale (Exit)			0	31-Mar-23				

Figure 10.42 Monte Carlo Simulation 2.

For the office yield, we will enter the following formula:

= NORMINV(RAND(),MEAN,STANDARD DEVIATION)

	B	C	D	E	F	G	H	I	J
19	REVENUES			NIA	Rent	Net Yield	Gross	Monte Carlo	Base Line
20				ft2	ft2	%	Value		
21	Office Units			5,400	50.00	4.00%	6,461,751	=NORMINV(RAND(),0,1%)	4.18%
22	Retail Units			1,000	80.00	6.00%	1,333,333	0.00%	6.00%
23									
24	GROSS DEVELOPMENT VALUE (GDV)						7,795,085		

Figure 10.43 Monte Carlo Simulation 3.

There is no need to ROUND this number because the yield can take any value. This type of data is called *continuous*.

And the baseline will become:

	B	C	D	E	F	G	H	I	J
19	REVENUES			NIA	Rent	Net Yield	Gross	Monte Carlo	Base Line
20				ft2	ft2	%	Value		
21	Office Units			5,400	50.00	4.00%	6,461,751	0.18%	=G21+I21
22	Retail Units			1,000	80.00	6.00%	1,333,333	0.00%	6.00%
23									
24	GROSS DEVELOPMENT VALUE (GDV)						7,795,085		

Figure 10.44 Monte Carlo Simulation 4.

For the retail yield, we could do the same thing, i.e., using a NORMINV formula. However, using a totally different NORMINV formula for the retail yield will mean that it may not make sense in relation to the office yield, as we know they are both highly correlated given that they are the same asset class and will be part of the same property in this analysis.

We will infer that the retail yield has a 1.2 correlation with the office yield. Therefore, the formula for the retail yield will be:

	B	C	D	E	F	G	H	I	J
16									
17	VALUATION								
18									
19	REVENUES			NIA	Rent	Net Yield	Gross	Monte Carlo	Base Line
20				ft2	ft2	%	Value		
21	Office Units			5,400	50.00	4.00%	6,587,170	0.10%	4.10%
22	Retail Units			1,000	80.00	6.00%	1,307,477	=I21*1.2	6.12%
23									
24	GROSS DEVELOPMENT VALUE (GDV)						7,894,647		

Figure 10.45 Monte Carlo Simulation 5.

Then, the baseline for the retail units will be the same formula as J21:

	B	C	D	E	F	G	H	I	J
16									
17	VALUATION								
18									
19	REVENUES			NIA	Rent	Net Yield	Gross	Monte Carlo	Base Line
20				ft2	ft2	%	Value		
21	Office Units			5,400	50.00	4.00%	5,282,650	1.11%	5.11%
22	Retail Units			1,000	80.00	6.00%	1,090,916	1.33%	=G22+I22
23									
24	GROSS DEVELOPMENT VALUE (GDV)						6,373,566		

Figure 10.46 Monte Carlo Simulation 6.

216 Real Estate Financial Modelling in Excel

If you press F9 several times, you will see the values changing each time. At this time, you can delete the sensitivity and scenario calculations if you want (as long as you have saved the initial file as a.xlsx and know where that file is stored. If not, do it now!).

Automating the Returns Calculations

As we mentioned before, to create a Monte Carlo simulation, you will need to run the returns many, many, many times. Some theorists say at least 1,000 times for the simulation to be valid. We would then need to press F9 1,000 times and record this number as a value 1,000. If you want to apply this technique – good luck! It will take a long time to finish it though. Instead, we can automate the process using **Macros.**

For the macro:
To create a macro, we will need the **Developer** tab in your Excel.

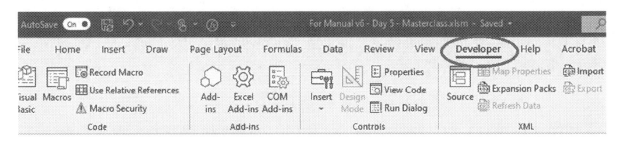

Figure 10.47 Monte Carlo Simulation 7.

If you have never used Macros, you will need to 'activate' it using the following steps:
File > Options > Customize the Ribbon
Tick the **Developer** box.

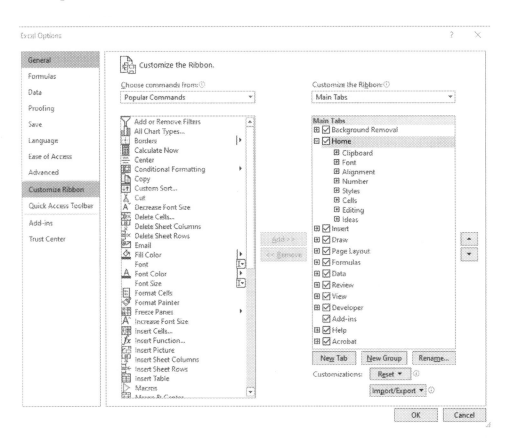

Figure 10.48 Monte Carlo Simulation 8.

And then **OK**.

You will now be able to see the **Developer** tab in the Ribbon.

Then, to make Excel calculate and copy and paste the returns a thousand times automatically, we will need to code the command for it.

I create a new sheet called MCarlo_IRR. This sheet will store the numbers from the Monte Carlo.

Then, I go to programming the macro. The quickest way I find to deal with macros is to start recording the macro doing what I want automated and then 'adjust' the command in Visual Basic thereafter. So, I go:

Developer > Record Macro

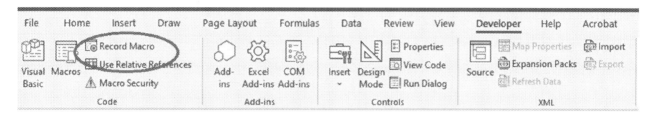

Figure 10.49 Monte Carlo Simulation 9.

Then:

Step 1: Name the macro and then OK (I named it MCarlo_IRR, but you can choose any other names).

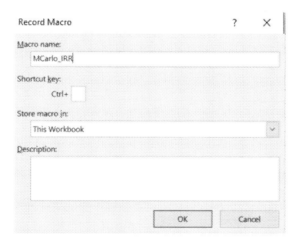

Figure 10.50 Monte Carlo Simulation 10.

Step 2: Press **F9** once.
Step 3: Click on the cell for the IRR and press **CTRL + C**.
Step 4: In the Monte Carlos sheet, **copy and paste special value**.
Step 5: **Stop Recording**.
Step 6: Click on **Macros**.
Step 7: When you see the name of your macro, select that macro and click on **Edit**.
Step 8: You will now need to create the 'loop'. The highlighted cells are the commands that will make your macro calculate, copy, and paste the many IRRs for the Monte Carlo (or any other numbers in your workbook for that matter).

```
Sub MCarlo_IRR() '
' MCarlo_IRR Macro '
For x = 1 To 100
Sheets("Sheet where your Monte Carlo random variables have been calculated").Select
Range("IRR Calc Cell").Select
Calculate
Selection.Copy
Sheets("Sheet where IRRs will be pasted").Select
Range("IRR Cell List").Offset(x, 0).Select
Selection.PasteSpecial Paste:=xlPasteValues
Next
End Sub
```

Note that I will be running only 100 IRRs because the purpose here is to explain the process, and although macros can make our lives more efficient, it still takes some time to calculate 1000s of calculations. If you haven't deleted the data tables (sensitivity and scenario), this calculation will take even longer, so I am just doing 100. If you want to calculate more than 100, just change the code 'For x= 1 to 100' for 'For x=1 to 100000000' (I would not recommend it though as your computer may crash! 1,000 should be enough).

Before you run the macro, **save the file**. It is common that macros break models, so you want to avoid that by having at least an intact version before running the macro.

Now, to run the macro, click on **Developer > Macros >** Select the macro you just created > Click **Run**.

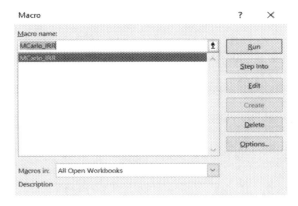

Figure 10.51 Monte Carlo Simulation 11.

Analysing the Result

To finalise, we will need to analyse the output; in our case, the 100 IRRs. To do so, we will use the Statistical Analysis ToolPak which is an **Add-in** (where you find **SOLVER**).

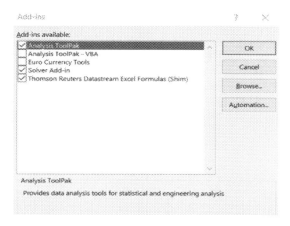

Figure 10.52 Monte Carlo Simulation 12.

Go to **Data > Data Analysis > Descriptive Statistics**

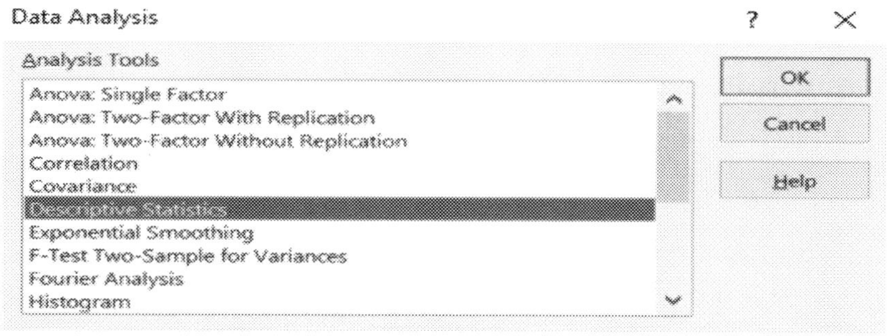

Figure 10.53 Monte Carlo Simulation 13.

Click **OK**.
Then, fill out:
Input Range: your 100 IRRs (tick the 'Labels in first row' if relevant)
Output Options: where you want the output data to be inserted.
I chose **Output Range** and cell just next to the IRR series.
Summary Statistics: tick that because that's what we need.
Click **OK**.

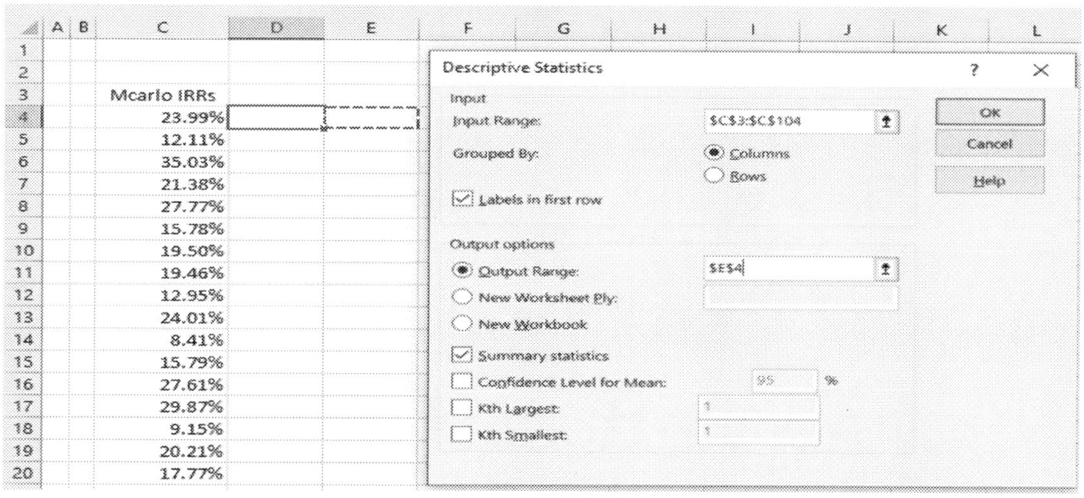

Figure 10.54 Monte Carlo Simulation 14.

The results will be:

	A	B	C	D	E	F	G	H
3			Mcarlo IRRs					
4			23.99%			Mcarlo IRRs		
5			12.11%					
6			35.03%		Mean	24%		
7			21.38%		Standard Error	1%		
8			27.77%		Median	23%		
9			15.78%		Mode	#N/A		
10			19.50%		Standard Deviation	11%		
11			19.46%		Sample Variance	1%		
12			12.95%		Kurtosis	202%		
13			24.01%		Skewness	93%		
14			8.41%		Range	65%		
15			15.79%		Minimum	3%		
16			27.61%		Maximum	69%		
17			29.87%		Sum	2443%		
18			9.15%		Count	101		

Figure 10.55 Monte Carlo Simulation 15.

Histogram

We can now create a histogram, i.e., a chart with the probabilities of IRRs. We will do so still using the Data Analysis add-in, but now, adding the Histogram:

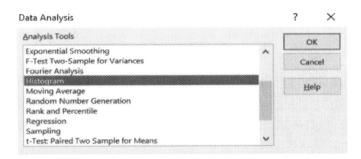

Figure 10.56 Monte Carlo Simulation 16.

Fill out the form:

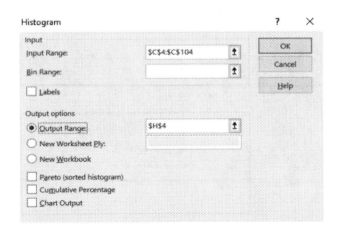

Figure 10.57 Monte Carlo Simulation 17.

Input: This is the series of IRRs you have just calculated using macros.

Bin range: If you don't add anything, Excel can calculate the best bins (the ranges of IRRs) for you.

Output range: Select a cell that is adjacent to your lists; you will then have everything close together for easier visualisation.

Do not tick anything else (Pareto, Cumulative, Chart)

Your histogram will be added next to the descriptive statistics table:

	A	B	C	D	E	F	G	H	I
4			23.99%		Mcarlo IRRs			Bin	Frequency
5			12.11%					3%	1
6			35.03%		Mean	24%		10%	5
7			21.38%		Standard Error	1%		16%	15
8			27.77%		Median	23%		23%	30
9			15.78%		Mode	#N/A		29%	25
10			19.50%		Standard Deviation	11%		36%	11
11			19.46%		Sample Variance	1%		42%	6
12			12.95%		Kurtosis	202%		49%	6
13			24.01%		Skewness	93%		55%	1
14			8.41%		Range	65%		62%	0
15			15.79%		Minimum	3%			
16			27.61%		Maximum	69%			
17			29.87%		Sum	2443%			
18			9.15%		Count	101			

Figure 10.58 Monte Carlo Simulation 18.

Now, you can create a chart to illustrate the IRR distributions better:

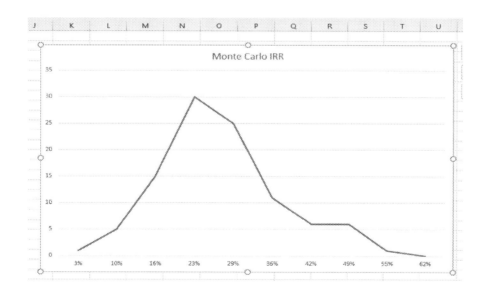

Figure 10.59 Monte Carlo Simulation 19.

Issues with Monte Carlo Simulation:

In my view, the only problem with Monte Carlo is assuming that we may over- or under-estimate the assumptions and probabilities. For example, we may assume that the market rent is 100 with a maximum of 150 and minimum of 50, and therefore, assume a standard deviation of 16.7, i.e.,

(150-50)/6. However, this market can be much more volatile than expected and therefore the standard deviation could be much higher. Another problem may be with correlations, for example, if we assume that the correlation between rental growth and yields are negative (higher rental growth, lower yields) at a rate of 0.5, i.e., for each 1% rental growth, the yield would be 0.5% lower, but because capital markets and the occupational markets may not be aligned overtime, historic correlations may not be applicable in future estimates. Moreover, variables which may be considered uncorrelated, weather and occupancy level, may indeed have a strong correlation which is not currently undertaken by the model.

Risks Analysis with Monte Carlo:

When we have competing properties for investment, Monte Carlo can give us a good visualisation of where risks lie in our portfolio. The chart can be an example of risk analysis that can be undertaken using Monte Carlo simulations.

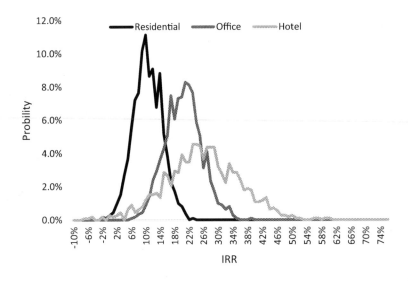

Figure 10.60 Monte Carlo Simulation 20.

Climate Risk Modelling

It is inevitable that when analysing risk, one should look at ESG (environmental, social, and governance) risks. In this book, we will look at the climate risk and how to assess this within the framework of financial risks.

From a financial modelling perspective, there are two main areas to consider: the transition risks and the physical risks, as they are classified by the TCFD (Task Force on Climate-Related Financial Disclosures).

As a risk, the best approach to deal with climate risk is through a **Monte Carlo simulation**, since we can use the terminology for flood risk in the simulation, i.e., return periods and annual probability.

For the transition risks, which are mainly related to market risks, these can be modelled implicitly through a greater standard deviation in exit yields, greater vacancy and rent-free period assumptions, and lower rental growth due to increased depreciation. If modelling debt, less climate-resilient properties may also face higher cost of borrowing.

For the physical risks, we can take the Representative Concentration Pathway (RCP) 4.5 which represents a medium scenario with global mean surface temperature change of 1.4 degrees Celsius and analyse the probability of inundation and the impact on insurance, excess cost from the insurance, and further climate-resilience CAPEX.

In our case, we assumed the costs as in the figure for flood risk assessment, i.e., insurance costs of £50,000 for a £5 million cover and insurance growth rate of 2.0% p.a.

	O	P	Q	R	S	T	U	V	W	X
9										
10	**Flood Risk Assessment**									
11										
12	Scenario: RCP4.5, 2030									
13	Insurance Cost	50,000								
14	Cover	5,000,000								
15	Base Growth	2.00%								
16										
17		Return	% Probability	Excess	Additional	Resilience	Expected	Stand,		
18	Inundation Level	Period	Annually	Costs	Insurance	Costs (£)	Period (qtrs)	Dev (qtrs)	Simulation 1	Simulation 2
19	0.05 ≥ inundation ≥ 0	1 in 5	20.00%	150,000	5%	1,250,000	20	5	26	37
20	0.25 ≥ inundation ≥ 0.05	1 in 100	1.00%	2,500,000	10%	5,000,000	400	100	583	1045
21	0.5 ≥ inundation ≥ 0.25	1 in 1,000	0.10%	25,000,000	15%	10,000,000	4,000	1,000	4985	8078
22	≥ 0.5	1 in 10,000	0.01%	100,000,000	20%	20,000,000	40,000	10,000	46203	64494

Figure 10.61 Climate Risk Modelling 1.

It is assumed that there is a 20% probability, or 1 in every 5 years that there will be an inundation level between 0 and 0.05 m, and if this happens, then there will be an excess cost of £150,000 and the insurance premium will increase by 5.0% in the following period. Furthermore, it is estimated that the property will have to incur an extra £1.25 million in resilience capital expenditure costs. Similarly, there is a 1 in 100 years event of inundation level greater than 0.05 m but less than 0.25 m. If this occurs, then the excess cost will be £2.5 million and an additional insurance of 10% will occur with further £5 million in resilience CAPEX to be incurred. The other inundation levels follow the same logic.

The simulation will be based on the NORMINV function as in the Monte Carlo simulation, with the following inputs:

=NORMINV(rand(), mean, standard deviation)
=NORMIV(rand(), expected period (i.e., the return period in quarters), std dev is estimated as a quarter of the mean)

As we are running a 10-year holding period and the first event can happen every 5 years, I have also added a second simulation if the investor happens to face two 0.05m inundation events during the life of the investment.

Since the simulation variable is the quarter in which the event may occur, this figure needs to be rounded to an integer so it can match the periods in the cash flow. Furthermore, this figure cannot be less than 0. The final formula will then be for Simulation 1:

=MAX(ROUND(NORMINV(RAND(),expected period, stand deviation),0),0)

For Simulation 2, the second event will happen *after* the first event; therefore, we will need to add the first event into the NORMINV function:

=MAX(ROUND(NORMINV(RAND(),expected period, stand deviation),0),0) + Simulation 1

We can then explicitly model climate insurance, climate excess costs, and climate resilience CAPEX in the cash flow.

Inundation Level	Return Period	% Probability Annually	Excess Costs	Additional Insurance	Resilience Costs (£)	Expected Period (qtrs)	Stand, Dev (qtrs)
Flood Risk Assessment							
Scenario: RCP4.5, 2030							
Insurance Cost	50000						
Cover	5000000						
Base Growth	0.02						
0.05 ≥ inundation ≥ 0	5	=1/P19	150000	0.05	1250000	=P19*4	=U19/4
0.25 ≥ inundation ≥ 0.05	100	=1/P20	2500000	0.1	5000000	=P20*4	=U20/4
0.5 ≥ inundation ≥ 0.25	1000	=1/P21	25000000	0.15	10000000	=P21*4	=U21/4
≥ 0.5	10000	=1/P22	100000000	0.2	20000000	=P22*4	=U22/4

Figure 10.62 Climate Risk Modelling 2.

Inundation Level	Stand, Dev (qtrs)	Simulation 1	Simulation 2
0.05 ≥ inundation ≥ 0	=U19/4	=MAX(ROUND(NORMINV(RAND(),$U19,$V19),0),0)	=MAX(ROUND(NORMINV(RAND(),$U19,$V19),0),0)+W19
0.25 ≥ inundation ≥ 0.05	=U20/4	=MAX(ROUND(NORMINV(RAND(),U20,V20),0),0)+W19	=MAX(ROUND(NORMINV(RAND(),$U20,$V20),0),0)+W20
0.5 ≥ inundation ≥ 0.25	=U21/4	=MAX(ROUND(NORMINV(RAND(),U21,V21),0),0)+W20	=MAX(ROUND(NORMINV(RAND(),$U21,$V21),0),0)+W21
≥ 0.5	=U22/4	=MAX(ROUND(NORMINV(RAND(),U22,V22),0),0)+W21	=MAX(ROUND(NORMINV(RAND(),$U22,$V22),0),0)+W22

Figure 10.63 Climate Risk Modelling 3.

Climate Insurance

I have named it as 'Climate Insurance', but this may well be part of the building insurance, and as such, part of the service charge. However, as the point was to model the impact of climate risk on insurance policies, I have made it an explicit cost.

The idea of the 'Climate Insurance' modelling is that if a 1 in 5 event occur, the premium will increase by 5%; if a 1 in 100 event occur, the premium will increase by 10%, and so on.

I start with the 'base cell' in Period 0 as = Insurance / 4 or £50,000 / 4 as the 'anchor cell'.

Then, I model from Period 1 onwards using the SUMPRODUCT:

=IF (SUMPRODUCT((Cash Flow Period = Simulation 1 Array) + (Cash Flow Period =Simulation 2 Array)),(1 + SUMPRODUCT(((Cash Flow Period = Simulation 1 Array) + (Cash Flow Period =Simulation 2 Array)) × (Additional Insurance Array))), (1 + (growth))^(Cash Flow Year – Previous Cash Flow Year))

In English, this formula means:

> If the quarter that you are in the cash flow equals a quarter when the simulation predicts a flood event to occur, then refer to the additional insurance table because less common events will have a higher additional cost in your insurance premium. If the quarter that you are in is not a

period of simulated flood, then your insurance premium is the same as the previous period but increased by a growth rate per year, say the inflation rate.

Climate Excess Costs

The climate excess costs refer to the costs that go beyond the insurance cover. As such, in our example, the first event will have an excess cost of £150,000, whereas the second event will have an excess cost of £2.5 million and so on. I also assumed that this would grow at the same rate as the insurance growth rate.

Still using the SUMPRODUCT, we will have (no need for the 'anchor' cell now and this will be modelled from Period 0):

=-SUMPRODUCT ((((Cash Flow Period =Simulation 1 Array) + (Cash Flow Period = Simulation 2 Array)) × (Excess Costs Array)) × (1 + growth)^year

Climate Resilience CAPEX

It is expected that once the property has been hit by a flood, then further climate resilience CAPEX will need to take place. For example, raising the floor height to prevent against flooding, installing pumps in basement areas, etc.

From a modelling perspective, it is the same logic as the climate excess costs, only that it will refer to another array.

=-SUMPRODUCT ((((Cash Flow Period =Simulation 1 Array) + (Cash Flow Period = Simulation 2 Array)) × (**Climate Resilience** Array)) × (1+ growth)^year

It is important to note that if the owner is investing in climate resilience measures today, then the costs of further climate resilience in the future will be lower and the investor should be able to benefit from lower transition risks, i.e., improved perception in the market that the building is resilient and therefore should command a higher value.

AI	AO / AP	AQ	AR	AS	AT	AU	AV	AW
2	**PRO-FORMA CASH FLOW**							
3	Year	0	0	0	0	1	1	1
4	Period	0	1	2	3	4	5	6
5	Date	30-Sep-21	31-Dec-21	31-Mar-22	30-Jun-22	30-Sep-22	31-Dec-22	31-Mar-23
6	Revenue	3,390,603	3,383,573	3,921,723	3,924,483	3,924,483	3,924,483	3,942,218
7	Net Rent	3,390,603	3,383,573	3,921,723	3,924,483	3,924,483	3,924,483	3,942,218
9	Operating Expenses	(195,687)	(195,687)	(195,687)	(195,687)	(199,818)	(379,642)	(12,750)
10	Business Rates	(107,100)	(107,100)	(107,100)	(107,100)	(109,344)	(109,344)	-
11	Void Costs	(76,087)	(76,087)	(76,087)	(76,087)	(77,724)	(77,724)	-
12	Letting Fees	-	-	-	-	-	(179,824)	-
13	Climate Insurance	(12,500)	(12,500)	(12,500)	(12,500)	(12,750)	(12,750)	(12,750)
14	Climate Excess Costs	-	-	-	-	-	-	-
16	Net Operating Income (NOI)	3,194,917	3,187,887	3,726,037	3,728,797	3,724,665	3,544,841	3,929,468
18	Investment Cash Flow	(271,248,240)	-	-	-	-	-	-
19	Purchase Price	(266,452,102)						
20	Refurb Capex	-	-	-	-	-	-	-
21	Climate Resilience Capex	-	-	-	-	-	-	-
22	Sale Price	-						
23	Acquisition Costs	(4,796,138)						
24	Sale Costs	-	-	-	-	-	-	-
26	Net Cash Flow (NCF)	(268,053,324)	3,187,887	3,726,037	3,728,797	3,724,665	3,544,841	3,929,468

Figure 10.64 Climate Risk Modelling 4.

	AO	AP	AQ	AR
13	Climate Insurance		=-P13/4	=(AR5<$AE$37)*AQ13*IF(SUMPRODUCT((AR4=$W$19:$W$22)+(AR4=$X$19:$X$22)), (1+SUMPRODUCT(((AR4=$W$19:$W$22)+(AR4=$X$19:$X$22))*($S$19:$S$22))), (1+($P$15))^(AR3-AQ3))
14	Climate Excess Costs		=-SUMPRODUCT(((AQ4=$W$19:$W$22)+ (AQ4=$X$19:$X$22))* ($R$19:$R$22))*(1+$P$15)^AQ3	=-SUMPRODUCT(((AQ4=$W$19:$W$22)+ (AQ4=$X$19:$X$22))* ($R$19:$R$22))*(1+$P$15)^AQ4
21	Climate Resilience Capex		=-SUMPRODUCT(((AQ4=$W$19:$W$22)+ (AQ4=$X$19:$X$22))* ($T$19:$T$22))*(1+$P$15)^AQ3	=-SUMPRODUCT(((AR4=$W$19:$W$22)+ (AR4=$X$19:$X$22))* ($T$19:$T$22))*(1+$P$15)^AR3

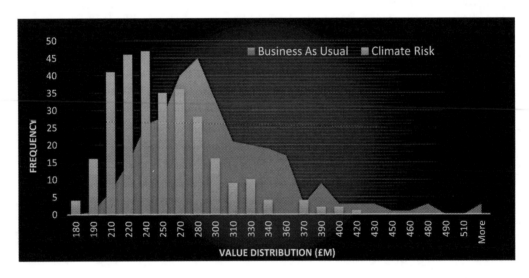

Figure 10.65 Climate Risk Modelling 5.

Now that you have incorporated climate risk calculations into your cash flow, the next step is to run the simulation with macros as described in the 'Monte Carlo Simulation' chapter.

Figure 10.66 Climate Risk Modelling 6.

The valuation distribution may look like the chart shown whereby 'business as usual' (area chart) may be underestimating the impact of climate risk (bar chart) on the value of the property.

Conclusion

When analysing risk in real estate, it is important to look at the sources of risks: macro and micro-economics, property, and market specific. Both qualitative and quantitative analysis of risk are important for assessing the quality of property project and both are subjective. However, when it comes to finance, a number will always be important, and in this case, you should focus on how all that was discussed in this section impact the ability of the property to generate cash via a discounted cash flow analysis, as we have learnt and discussed in previous sessions. With the help of data tables and visualisation tools, such as charts and conditional formatting, we can also quickly check if the numbers make sense and risk is acceptable given the estimated returns.

Final Words

Knowing everything about real estate financial modelling and structuring them efficiently and elegantly is quite a difficult task. As a final word, I want to leave you with some recommendations to master the art of financial modelling and become a proficient real estate investment analyst:

- **Practice is key**: keep modelling and follow the best practice guidance in this book for efficiency.
- **Understand the business case before using or developing models**: your inputs are fundamental to the final decision.
- **Use the financial models that accompany** this book to check your results and understanding. They can be found at www.cambridgerefinance.com.
- **Refer to this book** as a basis to build your own financial models and expand on them. Don't limit yourself to what is written here or what other people tell you: develop your own critical thinking.
- **Ask the experts**: If you admire anyone for their skills, don't be shy and ask them. It serves for financial modelling and anything else in life, really.

Index

Note: *Italic* page numbers refer to figures.

agency fees 111
all-risks yield method 29, 30, *30*
amortisation 110, 118; and amort 116; constant *see* constant amortisation; formula 136, *136*, 140, *141*; period 110
amortising loans 125–126, 128
annual cash flow 44, 50, 52, 62, 99–100
arrays 5, 83, 86, 89
asset cash flow 119, *119*, 185

bank rate 143, 144, *144*
Baum, A. 67
benchmarking 4
bootstrapping 144
breaks 4, 8, 14
budgeting 4

capital expenditure (CAPEX) *84*, 84–85
capital value 196, 198, 199, 203, *203*, 204
cash flow 46, *46*, *47*, *49*; annual 44, 50, 52, 62, 99–100; asset 119, *119*, 185; constant amount debt 131, *131*; dating 73–75; debt 118, 131, 137, 139, 188; discounted *see* discounted cash flow (DCF); indexation to inflation *60*, 60–61, *61*; investment 90, 92; JV partners' 192, *192*, *193*; net *see* net cash flow (NCF); NPV of 176, *177*; pro-forma development *see* pro-forma development cash flow; reserve 161, *162*; review cycles *57*, 57–58, *58*; single-let 44; single rent review or renewal to ERV *53*, 54–55, *55*; waterfall – equity first 185
cash release 161
cash reserve modelling: cell names *160*; interest income 160; labels & formulas 161; layout *160*; minimum balance required 160; minimum DSCR 160; minimum ICR 160; principal 159
cash sweep 138–139, *139*; amortisation formula 140, *141*; debt structure modelling *139*, 139–140, *140*; interest payment formula (MIN and MAX functions) 140

cash trap 160, 161
cell names 114, 115, 129, *160*
CHOOSE function 149–150, *150*, *151*
climate excess costs 224, 225
climate insurance 224–225
climate resilience CAPEX 224–226
climate risk modelling 222–224, *223–226*
commitment fees 111, 162
conditional formatting 7, 108, *109*, *110*, *112*, 204, 204–207, 226
constant amortisation: changing cell references 130, *130*; constant amount debt cash flow 131, *131*; copying and pasting to CA debt schedule 129, *130*; debt structure modelling 128–130; fixing cells from I/O debt structure 129, *129*; fully amortising 126, *126*, *127*; input data 128, *128*, *129*; renaming cells 129
constant amortisation (%) formula 130, *131*
constant amount, partially amortising 126–127
constant payment (CP): debt structure 132, *132*, *133*; full amortisation 131, *132*
constant payment – balloon (CP-B) 134, *134*; amortisation formula 136, *136*; debt structure modelling 135, *135*
constant percentage (CA): partially amortising 127–128, *128*
cost curves 173
covenants: calculation *122*; debt *see* debt covenants; debt yield 122; DSCR 121; estimated values 122; formulas *122*; and geared returns 152–155; ICR 121; live 152–154, *152–154*; LTV 121; and margins 145
credit enhancements: cash reserve modelling 159–161; commitment fee 162; debt service reserve account (cash reserve) 159; expiry 163; interest + fees paid 163, *163*, *164*; layout *163*; loan drawdown 163; principal repayment 163; revolving facility modelling 162; upfront fee 162; utilisation fee 162
Crosby, N. 67

Data Table function 196
DATE function 74
DATEDIF function 21
dating cash flow 73–75
dating problems 72, 101
debt brought forward (B/F) 116
debt carried forward (C/F) 116, 118
debt cash flow 118, 131, 137, 139, 188
debt covenants: debt service coverage ratio 112; debt yield 113; interest coverage ratio 111–112; loan to value 111
debt drawdown 117, 118, 120
debt service coverage ratio (DSCR) 4, 111, 112, 121, 128, 132, 134, 137, 139, 140, 145, 155, 160, 161
debt service reserve account (cash reserve) 159, 162
debt structures: amortisation 110; amortisation and amort 116; amortisation period 110; amortising loans 125–126; and analysing results, comparing 149–164; annuity calculation – PMT 133; cash sweep 138–139, *139*; CHOOSE function 149–150, *150, 151*; constant amortisation 126, *126, 127*, 128–131; constant amount 126–127; constant payment 131–132; Constant Payment – Balloon *134*, 134–136; constant percentage 127–128, *128*; covenants and geared returns 152–155; credit enhancements 159–164; debt covenants 111–113; description and framework 106; fees 110–111; fill without formatting 151, *151*; floating interest rates 141–142; formatting *152*; geared *vs.* ungeared returns 156–158; input headers for 106, *106*; input table *113*, 113–114, *114*; interest only 118, *118*; interest rate type 106–107; maturity period 110; modelling 116, *116*, 128, 137, *139*, 139–140, *140*, 145–148, *145–147*; rolled-up interest 136–137, *137*; types 106
debt yield 111, 113, 119, 122, 128, 132, 134, 137, 139, 140, 145, 150, 155
decision making 4, 14
development costs 170; construction period and development status 172, *172, 173*; distribution 173–175; hard costs 170–171, *171, 175*; known-curve 175; S-curve 174, *175*; site costs 170, *170*; soft costs 171, *172*; straight line 173–174, *174*; Z-score 174
development debt structures 187–188; effective senior debt cost 188; senior debt 187, *187, 188*
development funding financial modelling 186–193; development debt structures 187–188; effective senior debt cost 188; equity and debt draws 186–187; geared development returns and JV partners' returns 193, *193*; joint-venture – promote waterfall 190–191, *191*; JV partners' cash flows 192, *192, 193*; mezzanine debt modelling *189*, 189–190; mezzanine finance 189; notional finance costs 190; preferred returns cash flow 192, *192*; senior debt 187, *187, 188*; total debt finance cost 190–191
development revenue inputs 169, *169*
development time: as critical risk element 167–168, *168*; development costs 170; development costs distribution 173–175; development phases and durations input 168, *168*; goal seek *vs.* solver 182–183; pro-forma development cash flow 176, *176*
discounted cash flow (DCF) 17, 167; annual cash flow 44; arrays 83; CAPEX *84*, 84–85; estimated rental value 64; fixed rent over holding period 44–51; growth series *75*, 75–76; implicit valuation with 51–52; indexation to inflation 59–61; investment 62; London SW1, Office Prime West End Investment (case study) 94–97, *95–100*; market value 97–98; multi-let cash flows 65–70; net operating income 90; OPEX 85–89; passing rent 77–78, *78*; pro-forma cash flow 89–90; quarterly discounted cash flow 52; returns calculation 92–94; revenue 90–92; review cycles 56–59; review to market rent 79–81, *80–84*; single-let cash flows 44; single rent review or renewal to ERV 52–56; summarised annual cash flow 99–100; tenancy schedule *70*, 70–75; total rent forecast 83; void, rent-free period, and second lease 81–82
discount rate 46, 50–53, 53, 55, 69, 79, 93, 97, 98, 177, 178, 182

EDATE function 47–48, 54, 80, 82
equity and debt draws 186–187
equity multiple (EM) 56, 71, 92, 94, 164, 177, 194
equity/peak equity 94
equivalent yield 11, 12, 16, 19, 28, 30–35, *32–34*, 42
estimated rental value (ERV) 12, 13, 17, 19, 26, 28–30, 44–46, *45*, 52–54, *53*, 57, 58, 60, 64, 68, 75, 77, 79–84, *82–83*, 87–89, 101
Excel 3, 17; auditable and transparent 5; Data Table function 196; DATEDIF function 21; errors 5; flexible 5; Goal Seek function 32, 164; IF function 54, 79, 117; OR function 57; PMT function in 133; slowness 5; SOLVER function 154; SUMPRODUCT function 31; syntax in 49–50, 60; virtually free 5
exit value (sale price) 11, 60, 65, 68, 91, 101
expiries/voids 14
extension fees 111

facility fee 111
fees: commitment 162; debt structures 110–111; due 117, 120; interest + fees paid 163, *163, 164*; letting 71, *87*, 87–88, *88*, 90, 101; paid

117, 118, 120, 141; review 89, *89*; upfront 162; utilisation 162

financial modelling: benchmarking 4; budgeting and planning 4; creating layout (phase 2) 6; decision making 4; developing formulas (phase 3) 6; do-cumentation (phase 4) 7–8; financing 3–4; golden rules for 8–9; investment analysis 3; market analysis 4; revised model structure (phase 4) 6–7; rigorously testing model (phase 4) 7; risk assessment 3; spreadsheets 5; strategic asset management 4; transparency and communication 4; understand business case (phase 1) 6; uses of 3–4; valuation 4

financial modelling, golden rules for: add sensitivity tables (Rule No. 5) 8–9; never use hardcoded values in any formulas (Rule No. 4) 8; only one formula per row or column (Rule No. 2) 8; simple and elegant (Rule No. 1) 8; use intermediate calculations (Rule No. 3) 8; *see also* financial modelling

fixed rent over holding period 44; complete model 50; date and EDATE function 47–48; discount rate 46; ERV/market rent 45; exit yield 46; fixing cell reference *48*; holding period 45; initial yield 46; inputs 45–49; layout 44; outputs 49–51; output table 50; passing rent 45; period 46; purchase price 46, 48; rent 48; rental growth 45–46; review period 45; sale price 46, 48

fixing cell reference *48*, 76

floating interest rates 141–142; bank rate 143; covenants and margins 145; debt structure modelling *145*, 145–148, *146, 147*; forecasted bank rate 144, *144*; forward curve 144–145, *145*; interest rates forecast 144; short-term interest rates 143, *143*; yield curve 142

floating - margin 107

forecasted bank rate 144, *144*

formatting 35–41, *152*; borders 37, *37*; changing input and output cells colours 38–40, *38–40*; dates 36; hiding the gridlines 40, *40, 41*; passing rent, YP and values 37; valuation term 36; yield 37

forward curve 66, *66*, 144–145, *145*

froth 25, 79–81

geared development investment: capital structure of development projects 185; cash flow waterfall – equity first 185; commercial real estate debt and development activity 184–185; development funding financial modelling 186–193; geared vs. ungeared development returns 194; senior debt finance 185; total development costs and sources of funds 186; vs. ungeared development returns 194

geared development returns 177, 178, 193, 193, 194

geared *vs.* ungeared returns: linking chosen to live returns 156–157, *157, 158*; output table 158; sensitivity analysis 158, *159*

Goal Seek function 32, 164

goal seek *vs.* solver 182–183

graphs 200–202, *200–202*

gross development value (GDV) 166–169, *169*, 183, 184, 190

gross rent 13, 90

growth series *75*, 75–76

hard costs 170–171, *171*, 174, *175*

headline rent 13, 15, 79

histogram 220–222

holding period 45; fixed rent over *see* fixed rent over holding period; rental growth 68–69; revenue 91, *92*; single rent review or renewal to ERV 53

IF function 5, 8, 54, 74, 77, 79, 81, 88, 94, 121, 140, 146

indexation to inflation 59; cash flow *60*, 60–61, *61*; first review and indexation cycle 60; inputs 60; layout 59; output table *61*; sale price 60

INDEX function 152

INDEX MATCH function 157

indexed rents 14, 59

input headers 106, *106*

interest coverage ratio (ICR) 111–112, 119, 121, 122, 128, 132, 134, 137, 139, 140, 145, 155, 160, 161

interest due 112, 117, 120, 121, 140, 146

interest + fees paid 163, *163, 164*

interest income 160, 161

interest only (I/O) 118, *118*; asset cash flow 119, *119*; covenants 121–122; debt covenants 122–123; formulas 120, *120*; leverage improving financial returns 123–124; outputs calculations 124; structure – complete cash flow formulas *125*

interest paid 117, 118, 120, 123, 133, 137, 140

interest payment formula (MIN and MAX functions) 140

interest rates forecast 144

interest rate type 106–107; fixed – interest rate 107; floating - margin 107; loan to value 107

internal rate of return (IRR) 43, 49–50, *50, 51*, 166, 167, 194, 217–221; residual value on 181, *182*; sensitivity analysis 199, *199, 200*; single rent review or renewal to ERV 55; *vs.* XIRR 93

investment 62; analysis 3; cash flow 90, 92

known-curve 173, 175

labels and formulas 116, 161; amortisation 118; calls 161; cash inflow 161; cash outflow 161; debt brought forward 116; debt carried forward 118; debt cash flow 118; debt drawdown 117; fees due 117; fees paid 117; final capital repayment 118; interest due 117; interest

income 161; interest paid 117; net cash flow 161; reserve B/F 161; reserve cash flow 161, 162; reserve C/F 161; withdrawals 161
layer method 17, 18, 27
lease events: breaks 14; expiries/voids 14; indexed rents 14; lease starts 13; rent-free periods 13; rent reviews/renewals 13–14; step-ups 14
lease starts 13, 73, 83
letting fees 71, *87*, 87–88, *88*, 90, 101
live covenants 152–154, *152–154*
live returns: calculating 152–154, *152–154*; chosen to 156–157, *157, 158*
loan drawdown 163
loan to value (LTV) 107, 111, 154–155, *154–156*
LOOKUP function 76, 87, 172
LTC 190, *199, 200*, 203, *203, 204*

market analysis 4
market rent 12–13, 45, 53; first review to 79–80, *80*; second review to 80–81, *81*; third review to 81, *82–84*
market value 68, 93, 97–98
maturity period 110, 120, 136
mezzanine debt modelling *189*, 189–190
mezzanine finance 185, 189, 194
money market rate or treasury bill rate 143
Monte Carlo simulation 213, *214–222*, 215–218; automating returns calculations 216; randomising variables 214; risks analysis with 222
multi-let cash flows 65; global inputs 65; initial and exit yields 65–66; model structure 65; rental growth 67–69; risk premium 67

naming cells 114–115, *115*
net cash flow (NCF) 48, 54, 62, 90, 91, 117, 138, 161
net development value (NDV) 167–169, *169*
net operating income (NOI) 13, 89, 90, 111–113, 116
net present value (NPV) 43, 50, *50, 51*, 55, 166, 167, 196, 199, *199, 200*, 204; of cash flows 176, *177*; sensitivity analysis 199, *199, 200*; single rent review or renewal to ERV 55; *vs.* XNPV 93
net rents 10, 13, 22, 24, 90, 113, 168
NORMINV function 223

1-way data table 196–198, *197–199*
operating expenses 4, 22, 71, 90–92, *92*, 101, 196
OPEX 85–89; letting fees *87*, 87–88, *88*; review fees 89, *89*; void costs 85–87, *86*
OR function 57–59
origination fees (or front-end fees) 111, 162

passing rent 12, 37, 45, 77–78, *78*; lease end 77; lease start 77
periods 8, 13, 14, 46, 48, 52–54, 54, 57, 68, 69, 79, 85, 94, 112, 121, 122, 172, 174, 214, 222, 223

PMT: annuity calculation 133; calculating amortisation amount 133, *133, 134*
preferred returns cash flow 192, *192*
premium/default spread 145
prepayment fees 111
principal repayment 163
pro-forma development cash flow 71–72, *72*, 89–90, 176, *176*; NPV of cash flows 176, *177*; residual value on IRR 181, *182*; residual value on profit on cost 179; Solver function *180*, 180–181, *181*, 182; ungeared development returns 177–178, *178*; What-If Analysis 179, *179*
property yields (cap rates) 11
purchase price 12, 46, 48, 50, 54–56, 62, 68, 91, 108, 116, 117, 170
purchaser's costs 12, 35, 68, 91, 116, 117, 168–170

quarterly discounted cash flow 52

rack rented 29–30; all-risks yield method or true capitalisation method *29*; straight capitalisation method *29*
real estate valuation: all-risks yield method 30, *30*; capitalisation or implicit or traditional or conventional methods 17; equivalent yield 11–12; exit yield (or going-out cap rate) 11; initial yield (or going-in cap rate) 11; lease events 13–14; net *vs.* gross yields 12; output yields 30–34; property yields (or cap rates) 11; rack rented 29–30; rents 12–13; reversionary yield 11; under-rented scenario 17–24
'Red Book' Valuation 26
redemption fees 111
refurbishment CAPEX 91
renaming cells 129
rental growth 45–46, 53, 67–69; exit value (sale price) 68; holding period 68–69; initial value (purchase price) 68; target rate/discount rate 69, *70*; WACC 69
rent-free periods 12–14, 20, 79, 81–83, 90, 101, 169, 222
rent reviews/renewals 13–14
rents 48, 54; forecast 71, *71*; gross rent 13; headline rent 13; market rent/ERV/MRV 12–13; net rent 13; passing rent 12; topped-up passing rent 13
residual value: on IRR 181, *182*; on profit on cost 179
return on equity (ROE) 94
returns calculation: equity/peak equity 94; *IRR vs. XIRR* 93; *NPV vs. XNPV* 93; output table *94*; price *vs.* value *vs.* worth 93; profit 93; ROE and equity multiple 94
revenue: holding period 91, *92*; investment cash flow 90; net cash flow 91; net operating income 90; net rents 90; operating expenses

90; operating expenses and investment cash flow 92; purchase price 91; purchaser's costs 91; refurbishment CAPEX 91; revenue holding period switch 91; sale price/exit value 91
revenue holding period switch 91
review cycles 56–59, *57*; cash flow *57*, 57–58, *58*; complete model 59; layout 56; OR function 57–58; outputs table *58*; review periods 57
review fees 89, *89*
review periods 45, 53, 57–58, 60
risk premium 67
rolled-up interest 136–137, *137*; debt structure modelling 137, *138*; payment formula 138, *138*
ROUND function 214
Royal Institution of Chartered Surveyors (RICS) 17, 44

sale price 11, 46, 48, 54, 60, 65, 68, 91
sales proceeds 168–169, *169*, 177
scenario analysis 3, 167, 208–213, *208–213*
S-curve 174, *175*
senior debt finance 148, 185
sensitivity analysis 158, *159*; climate excess costs 225; climate insurance 224–225; climate resilience CAPEX 225–226; climate risk modelling 222–224, *223–226*; conditional formatting 204, *204–207*; finding IRRs and NPVs 199, *199*, *200*; graphs 200–202, *200–202*; notes for 208; 1-way data table 196–198, *197–199*; scenario analysis 208–213, *208–213*; stress testing 213–222; two-way data tables 202–204, *203, 204*
17 Fleet Street (case study) 19, *19*; hardcore method 25, *25, 26*; layer method 27, *27*; over-rented scenario 26; Term and Reversion *vs.* Hardcore 26; Term & Reversion 19–24, *20–24*, 26, *27*; years purchase for 27, *27*
short-term interest rates 143, *143*
single-let cash flows 44
single rent review 52; cash flow *53*, 54–55, *55*; equity multiple 56; ERV/market rent 53; holding period 53; initial yield on worth 56; inputs 53; IRR 55; layout 52; NPV 55; outputs 55–56; profit 55–56; rental growth and discount rate 53; review period 53; table 53, 56; worth/investment value 55
site costs 166, 170, *170*
soft costs 171, *172*, 190
solver function *180*, 180–182, *181*
spreadsheets 3, 5–7, *20–26*, 28
straight capitalisation method *30*
strategic asset management 4
stress testing 213; analysing result 218–220; histogram 220–222; Monte Carlo simulation 213, *214–222*, 215–218; setting capital requirements 213

SUMIF function 94
SUMPRODUCT function 31

tenancy schedule *70*, 70–71; apportioning calculation *72*, 73; dating cash flow 73–75; dating problems 72; intermediate calculations 71; output 71; pro-forma cash flow 71–72, *72*; rents forecast 71, *71*
Term & Reversion method 17–18, *18*, 20–24, *20–24*; 17 Fleet Street (case study) 19–27; Retail Single-Let West End Investment, London W1 (case study) 28–29
topped-up passing rent 13
total debt finance cost 190–191; joint-venture – promote waterfall 190–191, *191*; notional finance costs 190
total rent forecast 83
true capitalisation method *29*
two-way data tables 202–204, *203, 204*

under-rented scenario 17; layer method 18; Term & Reversion method 17–18, *18*, 20–24, *20–24*; under-rented hardcore method 18, *18*; years purchase for *18*, 18–19
ungeared development returns 177–178, *178*
upfront fee 162
utilisation fee 111, 162

valuation 4; discounted cash flow 51–52; financial modelling 4; formatting 36; output yields 36; real estate *see* real estate valuation
values 37, 49–50; cash flow 125; estimated 122; hardcoded 8, 9
void 13, 14, 81–82; costs 71, 85–87, *86*, 90, 94; period 82, 101, 214

WACC 69
What-If Analysis 9, 153, 158, 179, *179*
worth/investment value 50, *50, 51*, 55

yields 37; analysis and net value *34*, 34–35, *35*; curve 142; equivalent 11–12; net *vs.* gross 12; property 11; reversionary 11
yields (initial and exit) 46, 56, 65; bond yield 10Y historic 66, *67*; forward curve 66, *66*; or going-in cap rate 11; or going-out cap rate 11; risk-free rate 65–66; on worth 56; yield curve *66*
yields, output 30–34; borders 37, *37*; changing colours 38–40, *38–40*; dates 36; equivalent yield 31–34, *32–34*; formatting 35–41; hiding the gridlines 40, *40, 41*; initial yield 31; lock sheet *41, 42*; passing rent 37; protecting sheet 41–42; valuation term 36; values 37; YP 37

Z-score 174

Printed in the United States